THEORY OF ALGEBRAIC INVARIANTS

THEORY OF ALGEBRAIC INVARIANTS

DAVID HILBERT

translated by Reinhard C. Laubenbacher
New Mexico State University
edited by Reinhard C. Laubenbacher and Bernd Sturmfels
with an introduction by Bernd Sturmfels
Cornell University

Published by the Press Syndicate of the University of Cambridge
The Pitt Building, Trumpington Street, Cambridge CB2 1RP
40 West 20th Street, New York, NY 10011-4211, USA
10 Stamford Road, Oakleigh, Melbourne 3166, Australia

First published 1993
Reprinted 1994

Printed in the United States of America

Library of Congress cataloguing in publication data

Hilbert, David, 1862–1943.

Theory of algebraic invariants / David Hilbert ; translated by Reinhard C. Laubenbacher ; edited and with an introduction by Bernd Sturmfels.

p. cm.

Includes bibliographical references.

ISBN 0-521-44457-8. – ISBN 0-521-44903-0 (pbk.)

1. Invariants, I. Sturmfels, Bernd. II. Title.
QA201.H716 1993
512′.5–dc20 93-15650
CIP

A catalog record of this book is available from the British Library.

ISBN 0-521-44457-8 hardback
ISBN 0-521-44903-0 paperpack

Contents

Preface

Around the turn of the century, the University of Göttingen was a Mecca for mathematicians and students from around the world, including the United States. Besides Hilbert, who held Gauss's chair, the other three chairs were held by Klein, Minkowski, and Runge. The visitors took back with them a large number of handwritten lecture notes, some of which eventually found their way into mathematics libraries of several U.S. universities. The present notes of Hilbert's 1897 course on invariant theory comprise 527 handwritten pages, taken by Hilbert's student Sophus Marxsen, who went on to write a dissertation in invariant theory. As part of the estate of John Henry Tanner they were given to the Cornell University Mathematics Library.

The project to translate these notes and make them more widely available was suggested by Keith Dennis. Without his initiative and continuing support this translation would not have been undertaken. Steve Rockey at the Cornell University Mathematics Library was also very helpful. Finally, many thanks are due to the late Eleanor York for her expert typing of the manuscript.

The photograph on the cover of Hilbert in his family surroundings is from the collection of Keith Dennis, who kindly permitted its reproduction. Thanks are also due to Stephen R. Singer for his expert reproduction of the less than perfect original.

Reinhard C. Laubenbacher

Introduction

In the summer semester 1897 David Hilbert gave an introductory course in invariant theory at the University of Göttingen. The present text is an English translation of the handwritten course notes taken by Hilbert's student Sophus Marxsen.

When Hilbert gave this course in 1897, his research in invariant theory had been completed. In particular, Hilbert's famous Finiteness Theorem had been proved and published in two striking papers (Hilbert 1890, 1893).* These papers changed the course of invariant theory dramatically, and they laid the foundation for modern commutative algebra. Thus 1897 was a perfect time for Hilbert to give an introduction to invariant theory, taking into account both the old approach of his predecessors and his new ideas. It is this bridge from nineteenth-century mathematics into twentieth-century mathematics which makes these course notes so special and distinguishes them from other treatments of invariant theory.

Hilbert's course is at a level accessible to graduate students in mathematics. Prerequisites include familiarity with linear algebra and the basics of ring theory and group theory. The text provides a self-contained introduction to classical invariant theory, and it will be of interest to anyone who wishes to study this subject. But we believe that this translation will also be valuable as a historical source for experts in contemporary invariant theory. Mathematicians and computer scientists who are interested in algorithmic aspects of invariant theory can read the present notes in parallel with Sturmfels (1993).

Let us attempt to answer the question, "What is invariant theory?" A nineteenth-century mathematician might have answered, "Invariant theory is *the* bridge between algebra and geometry." This point of view

* For English translations of Hilbert's research papers in invariant theory see M. Ackerman (1978).

found its most explicit expression in Felix Klein's Erlangen Programm (Klein 1872). Today invariant theory is often understood as a common branch of representation theory, algebraic geometry, commutative algebra, and algebraic combinatorics. Each of these four disciplines has roots in nineteenth-century invariant theory. This will become evident from Hilbert's course.

The standard references to today's invariant theory include, among others, Dieudonné and Carell (1971), Springer (1977), and Mumford and Fogarty (1982). In modern terms, the basic problem of invariant theory can be characterized as follows. Let V be a K-vector space on which a group G acts linearly. In the ring of polynomial functions $K[V]$ consider the subring $K[V]^G$ consisting of all polynomial functions on V which are invariant under the action of the group G. The basic problem is to describe the invariant ring $K[V]^G$. In particular, we would like to know whether $K[V]^G$ is finitely generated as a K-algebra and, if so, to give an algorithm for computing generators.

Hilbert essentially proved the following theorem (see Section II.2 of this text): If G is a reductive algebraic group, then $K[V]^G$ is finitely generated as a K-algebra. A minimal set of generators is called a full system of invariants. Recall that a group G is *reductive* if all its linear representations are direct sums of irreducible representations. After Hilbert's complete solution to this outstanding problem of invariant theory, the field was pronounced dead, only to be resurrected thirty years later by the work of Hermann Weyl (1939), in which invariant theory was developed for all classical Lie groups. Weyl developed invariant theory as a special instance of representation theory.

A famous problem, left open by both Hilbert and Weyl, asked whether the Finiteness Theorem continues to hold for every subgroup of the general linear group $GL_m(\mathbf{C})$. It is known as Hilbert's fourteenth problem. In 1959 Nagata answered this question by giving an example of a (nonreductive) group G acting linearly on a vector space V such that $K[V]^G$ is not finitely generated. Popov (1979) extended Nagata's work by proving the following remarkable converse to Hilbert's Finiteness Theorem: For every nonreductive algebraic group G there exists a finitely-generated reduced k-algebra A and a rational action of G on A by k-automorphisms such that A^G is not finitely generated. Another important result in today's invariant theory is the theorem of Hochster and Roberts (1974), which states that the invariant ring $k[V]^G$ is Cohen-Macaulay, provided K has characteristic zero and G is reductive.

Classical invariant theory is concerned with the following special case: the group G is the group $GL_m(\mathbf{C})$ of invertible complex $m \times m$-matrices, acting by linear substitution on a space V of m-ary forms, that is, homogeneous polynomials in m variables. In fact, in Hilbert's course most of the discussion centers around the $m = 2$ case of binary forms. For a complementary point of view on the invariant theory of binary forms we refer to the exposition of Kung and Rota (1984).

We now give a summary of Hilbert's course. In Section I.1 he introduces the space $V = S^n\mathbf{C}^m$ of m-ary n-forms, that is, homogeneous polynomials of degree n in m variables. It is proved that this complex vector space has dimension $\binom{n+m-1}{n}$. The group $GL_m(\mathbf{C})$ acts on $S^n\mathbf{C}^m$ by linear substitutions, which amounts to the nth symmetric power representation. Starting with Section I.2, Hilbert specializes to the case $m = 2$. It is proved that $GL_2(\mathbf{C})$ is generated by diagonal matrices $\begin{pmatrix} \kappa & 0 \\ 0 & \lambda \end{pmatrix}$, lower triangular matrices $\begin{pmatrix} 1 & 0 \\ \mu & 1 \end{pmatrix}$, and upper triangular matrices $\begin{pmatrix} 1 & \nu \\ 0 & 1 \end{pmatrix}$. The concepts of invariants and covariants are introduced in Section I.3.

Let $a_0, a_1, \ldots, a_n$ and x_1, x_2 be coordinates on $S^n\mathbf{C}^2$ and $\mathbf{C}^2$, respectively. A *covariant* is a polynomial $I(a_0, a_1, \ldots, a_n; x_1, x_2)$ which is fixed under the action of any matrix $T \in GL_2(\mathbf{C})$, up to a factor $\det(T)^p$. If I depends only on $a_0, a_1, \ldots, a_n$, then it is called an *invariant.* The number p is called the *weight* of the co- or invariant. It is extremely important to note that here $GL_2(\mathbf{C})$ acts on $\mathbf{C}^2$ by the contragredient (or inverse) representation (cf. Springer 1977, §3.1). Otherwise the whole concept of "covariants" does not make sense. The essence of this point is expressed in Hilbert's statement, "The simplest covariant is the form f itself," at the end of Section I.3.

The action of the group $GL_2(\mathbf{C})$ gives rise to an action of the Lie algebra $sl_2(\mathbf{C})$ on the space $S^n\mathbf{C}^2$ of binary n-forms. The two Lie algebra generators corresponding to the lower- and upper-triangular matrices are denoted $\mathbf{D}$ and $\mathbf{\Delta}$. In Section I.4 it is shown that invariants are annihilated by $\mathbf{D}$ and $\mathbf{\Delta}$. This characterizes invariants among all polynomials in $a_0, a_1, \ldots, a_n$ which are invariant under the action of the diagonal matrices only. A similar criterion is proved for covariants. In Section I.5 we see that in this criterion any one of the two conditions $\mathbf{D}I = 0$ and $\mathbf{\Delta}I = 0$ is implied by the other, and hence can be omitted. In order to show this, Hilbert proves the commutation relation

$$(\mathbf{D\Delta} - \mathbf{\Delta D})\mathcal{A} = (ng - 2p)\mathcal{A}$$

for the action of the Lie algebra generators on the space of polynomials in $a_0, \ldots, a_n$ of degree g and weight p. We refer to Serre (1987, Chapter IV) for an introduction to the representation theory of the Lie algebra $sl_2(\mathbf{C})$.

The action of the Lie algebra generators $\mathbf{D}$ and $\mathbf{\Delta}$ leads in Section I.6 to the "smallest system of conditions for the determination of invariants and covariants." Every covariant can be written as

$$\mathcal{C}(a_0, \ldots, a_n; x_1, x_2) \quad = \quad \sum_{i=0}^{m} C_i(a_0, \ldots, a_n)\, x_1^{m-i} x_2^i.$$

The leading coefficient C_0 is called the *source* of the covariant. The main result in Section I.6 is that the following system of differential equations characterizes covariants:

$$\mathbf{D}C_i \quad = \quad i\, C_{i-1} \qquad (i = 1, \ldots, m).$$

These equations imply in particular that each covariant is determined by its source.

Section I.7 is concerned with the Hilbert series of the rings of invariants and covariants. A self-contained introduction is given to the enumerative calculus of Cayley and Sylvester. It is similar to the presentation in Springer (1977). In Section I.8 the concept of *transvections* is introduced. It is proved that each covariant of a binary form f can be expressed as a polynomial in the transvections $f_1, \ldots, f_n$, divided by a suitable power of f itself. Several applications are given, including the determination of the complete system of covariants for the binary quartic. The natural generalization of the concept of invariants and covariants to several base forms is presented in Section I.9. In Section I.10 Hilbert introduces three procedures for generating new covariants from old ones. The first one is the remarkable statement: "Covariants of covariants are again covariants." The other two are the *polarization process* and the *Aronhold process*.

So far invariants and covariants were always represented explicitly as polynomials in $a_0, \ldots, a_n; x_1, x_2$. The last three sections of Part I are concerned with three alternative representations. In Section I.11 we think of (the roots of) a binary form of degree n as an unordered collection of n points on the complex projective line. This allows us to express each covariant as a symmetric function in the coordinates of these n points. The other two representations, in terms of one-sided derivatives (Section I.12) and using the symbolic method (Section I.13), are of historic interest, but they are less important for what follows in Part II of Hilbert's course.

The symmetric function representation of Section I.11 leads to a direct proof of Hilbert's Finiteness Theorem in the case of binary forms. This proof, which is also due to Hilbert, is given in Section II.1. See Sturmfels (1993, Section 3.7) for an alternative presentation of the same proof. The key lemma for this proof is the existence of a finite *Hilbert basis* for a system of linear equations over the natural numbers. Today this lemma is foundational for the theory of integer programming (Schrijver 1977). In Section II.1 of Hilbert's course we can fully appreciate the purpose for which "Hilbert bases" were invented.

The Finiteness Theorem for binary forms is due to Gordan (1868). Gordan's original proof gave an explicit construction of the complete system of invariants based on transvections and the symbolic method, but it is much more complicated than the one given in Section II.1. However, both Gordan's proof and Hilbert's proof in Section II.1 do not generalize to forms in $m \geq 3$ variables. The "generalizable proof" is presented in Section II.2. This is the proof which in 1890 shocked the invariant theory community. It is based on an averaging operator, called *Cayley's* Ω*-process*, and another key lemma, called *Hilbert's Basis Theorem* for polynomial ideals. This proof works in full generality, but it is not constructive.

The remaining sections of the course are based on Hilbert's 1893 paper "On the complete system of invariants." In this paper Hilbert essentially gives an explicit algorithm for computing a full system of invariants. In Section II.3 Hilbert discusses the relation between the ring of invariants and the field of invariant rational functions. He proves that the invariant ring is a unique factorization domain, and that it coincides with the intersection of the invariant field with the polynomial ring $\mathbf{C}[V]$.

The material presented in Section II.4 is fundamental both for Hilbert's algorithm and for geometric invariant theory (Mumford and Forgarty 1982). The vanishing locus of all homogeneous invariants is called the *nullcone*. The forms lying in the nullcone are called *nullforms*. The main theorem in Section II.4 states that, if $I_1, \ldots, I_k$ are any invariants which define the nullcone set-theoretically, then the invariant ring $\mathbf{C}[V]^G$ is integral over $\mathbf{C}[I_1, \ldots, I_k]$. The algorithmic significance of this result is explained in Sturmfels (1993, Section 4.6). The key lemma in the proof of this main theorem is *Hilbert's Nullstellensatz*. It is the very purpose for which the Nullstellensatz was first invented. Several applications of the main theorem to binary forms are given.

In Section II.5 Hilbert gives a more precise combinatorial description of the ternary nullforms. They are precisely those ternary forms f such that after a suitable linear transformation, the Newton polygon of f does not contain the origin. The geometric meaning of a ternary form being a nullform is that the planar curve $\{f = 0\}$ has certain types of singularities, which makes the curve *unstable.* This example is the point of departure in Mumford and Fogarty (1982). In geometric invariant theory the concept of stability (= lying outside the nullcone) is extended to arbitrary projective varieties, and, as the main application, certain moduli spaces are constructed.

Section II.6 deals with *Hilbert's Syzygy Theorem.* In modern language this theorem states that the polynomial ring $\mathbf{C}[x_1, \ldots, x_n]$ has *global dimension* n, that is, every module over the polynomial ring admits a finite free resolution of length at most n. The Syzygy Theorem is applied to prove the fact that the Hilbert series of the invariant ring is a rational function. In commutative algebra this is sometimes called the *Hilbert-Serre Theorem* (cf. Atiyah and Macdonald 1969, Theorem 11.1).

In the last three sections, II.7–9, Hilbert discusses applications of invariant theory to (algebraic) geometry, and he outlines future directions. Some of Hilbert's questions have received an answer during the past century (notably, Nagata's solution to Hilbert's fourteenth problem). Other classical questions are still open, and many new questions arise from the increasing demand for invariant theory as a tool in applied mathematics. These questions guarantee the future of invariant theory as an exciting area of research in mathematics.

References

Ackerman, M. (1978). *Hilbert's Invariant Theory Papers.* Math Sci Press, Brookline, Mass.

Atiyah, M. F., and Macdonald, I. G. (1969). *Introduction to Commutative Algebra.* Addison-Wesley, Reading, Mass.

Dieudonné, J., and Carell, J. B. (1971). *Invariant Theory – Old and New.* Academic Press, New York.

Gordan, P. (1868). Beweis, dass jede Covariante und Invariante einer binären Form eine ganze Funktion mit numerischen Coefficienten einer endlichen Anzahl solcher Formen ist. *J. Reine und Angewandte Mathematik* **69**, 323–54.

Hilbert, D. (1890). Über die Theorie der algebraischen Formen. *Math. Annalen* **36**, 473–531 (also: "Gesammelte Abhandlungen," II, 199–257, Springer, 1970).

Hilbert, D. (1893). Über die vollen Invariantensysteme. *Math. Annalen*

42, 313–70 (also: "Gesammelte Abhandlungen," II, 287–344, Springer, 1970).

Hochster, M., and Roberts, J. (1974). Rings of invariants of reductive groups acting on regular rings are Cohen-Macauley. *Advances in Math.* **13**, 115–75.

Klein, F. (1872). Vergleichende Betrachtungen über neue geometrische Forschungen. Programm zum Eintritt in die philosophische Fakultät und den Senat der Universität zu Erlangen. A Deichert, Erlangen.

Kung, J. P. S., and Rota, G.-C. (1984). The invariant theory of binary forms. *Bulletin Amer. Math. Soc.* **10**, 27–85.

Mumford, D., and Fogarty, F. (1982). *Geometric Invariant Theory* (2nd Edition). Springer, New York.

Nagata, M. (1959). On the 14th problem of Hilbert. *American J. Math.* **81**, 766–72.

Popov, V. L. (1979). Hilbert's theorem on invariants. *Soviet Math. Doklady* **249**, 551–5.

Schrijver, A. (1986). *Theory of Linear and Integer Programming.* Wiley-Interscience, Chichester.

Serre, J.-P. (1987). *Complex semisimple Lie algebras.* Springer, New York.

Springer, T. A. (1977). *Invariant Theory.* Lecture Notes in Mathematics **585**. Springer, Heidelberg.

Sturmfels, B. (1993). *Algorithms in Invariant Theory.* Springer-Verlag, Vienna, to appear.

Weyl, H. (1939). *The Classical Groups – Their Invariants and Representations.* Princeton University Press.

Bernd Sturmfels

Theory of Algebraic Invariants

Lectures by

Prof. Hilbert

prepared by

Sophus Marxsen
cand. math.

Göttingen
Summer Semester 1897

I
The elements of invariant theory

Lecture I (April 27, 1897)

The theory of algebraic invariants, with which we want to concern ourselves here, is a modern discipline. Its origins can be traced back to Cayley (1845), who used the term "hyperdeterminants" for functions possessing the invariant property. At present, we only mention as references the four main text books which all differ from each other and treat the subject from different perspectives. These are the following:

Clebsch. *Theorie der binären algebraischen Formen.* Leibzig 1872.

Salmon. *Modern Higher Algebra.* Dublin 1885. German trans. by Friedler. Leibzig 1885.

Faà di Bruno. *Théorie des formes binaires.* Turin 1876. German trans. by Walter. Leibzig 1881.

Gordan. *Vorlesungen über Invariantentheorie.* Leibzig 1885 (in particular vol. II).

One should add to these: **Franz Meyer**, *Bericht über den gegenwärtigen Stand der Invariantentheorie*, 1892. (In the "Berichte der deutschen Mathematiker Vereinigung.")

The necessary prerequisite for an understanding of the following is a knowledge of differentiation and of the basic theorems from the theory of determinants. The latter can be found in, for example, **Bältzer**, *Determinantentheorie*; **Hasse**, *Raumgeometrie*, Lecture 7; **Salmon**, *Modern Higher Algebra*, Lecture 1; **Serret**, *Algèbre Supérieure*, Tome II, Ch. IV (of course also in the German translations); **H. Weber**, *Algebra*, vol. I, Sect. II.

The books by Salmon and Faà di Bruno are the best introduction to invariant theory.

I.1 The forms

A sum of products of constants and variables will be called a *polynomial.* Thus, if c_{ikl} are constants, and x, y, z are variables, then

$$\mathcal{F}(x,y,z) = \sum_{i,k,l} c_{ikl} x^i y^k z^l, \qquad i,k,l = 0,1,2,3,4,5,$$

is a polynomial in three variables. The general form of a polynomial is

$$\mathcal{F}(x,y,z,\dots) = \sum_{i,k,l,\dots} c_{ikl\dots} x^i y^k z^l \cdots, \qquad i,k,l,\dots = 0,1\dots,n.$$

Each of the products, that is, each expression $c_{ikl\dots} x^i y^k z^l \cdots$, is called a *term* of the polynomial. Its characteristic number $n = i + k + l + \cdots$, that is, the sum of the exponents of the variables, is called the *order of the term.* The order of the term with the highest order is called the *order of the polynomial.*

We always want to think of a polynomial as ordered, by taking first the term of order zero—the constant—then the terms of order one, then those of order two, etc; we want to indicate this by using the notation:

$$\mathcal{F}(x,y,z,\dots) = [0] + [1] + [2] + \cdots + [n],$$

where n is the order of the polynomial.

The last term $[n]$ can not be identically zero, because otherwise the order would be smaller than n. On the other hand it is possible that all terms other than $[n]$ vanish, in which case the function is called a *homogeneous function* or a *form.* In this case, therefore, all terms have the same order.

In the following we will only consider forms. We want to introduce fixed notation for them, which we shall use throughout. The order of a form is always n (respectively, $\nu, N, \dots$); the number of variables in general discussions will be m, so the variables will always be denoted $x_1, x_2, \dots, x_m$.

It is no restriction to consider only forms, since we can always obtain a form from a polynomial and vice versa. Indeed, if we add a variable x_m to

$$\mathcal{F}(x_1, x_2, \dots, x_{m-1}) = [0] + [1] + [2] + \cdots + [n],$$

then

$$\Phi(x_1, x_2, \dots, x_m) = [0]x_m^n + [1]x_m^{n-1} + [2]x_m^{n-2} + \cdots + [n]$$

is a homogeneous function, from which we can easily reconstruct the original function by setting $x_m = 1$:

$$\Phi(x_1, x_2, \ldots, x_{m-1}, 1) = \mathcal{F}(x_1, x_2, \ldots, x_{m-1}).$$

Moreover, we have

$$x_m^n \mathcal{F}\left(\frac{x_1}{x_m}, \frac{x_2}{x_m}, \frac{x_3}{x_m}, \ldots, \frac{x_{m-1}}{x_m}\right) = \Phi(x_1, x_2, \ldots, x_m).$$

Hence, the theory of forms in m variables is essentially identical to the theory of general polynomials in $m-1$ variables. We will now derive two theorems which we will use frequently. Let

$$\Phi(x_1, x_2, \ldots, x_m) = \sum_{i,\ldots,s} c_{ik\ldots s} x_1^i x_2^k \cdots x_m^s, \qquad i + k + \cdots + s = n.$$

If we replace x_i by ux_i, where u is arbitrary, for example, a variable, then we obtain

$$\begin{aligned}\Phi(ux_1, \ldots, ux_m) &= \sum_{i,\ldots,s} c_{ik\ldots s} u^i x_1^i u^k x_2^k \cdots u^s x_m^s \\ &= \sum_{i,\ldots,s} c_{ik\ldots s} x_1^i x_2^k \cdots x_m^s u^{i+k+\cdots+s}.\end{aligned}$$

Therefore,

$$\Phi(ux_1, \ldots, ux_m) = u^n \sum_{i,\ldots,s} c_{ik\ldots s} x_1^i x_2^k \cdots x_m^s$$

or, finally,

$$\Phi(ux_1, \ldots, ux_m) = u^n \Phi(x_1, x_2, \ldots, x_m). \qquad (I)$$

A polynomial has this property precisely if it is a form, which can be seen by decomposing it into its homogeneous parts. It can consequently be used as a defining property of forms.

If we differentiate (I) with respect to u, which is admissible since it holds identically as an equation in u, then it follows that

$$\begin{aligned}&\frac{\partial\Phi(ux_1, \ldots)}{\partial(ux_1)} x_1 + \frac{\partial\Phi(ux_1, \ldots)}{\partial(ux_2)} x_2 + \cdots + \frac{\partial\Phi(ux_1, \ldots)}{\partial(ux_m)} x_m \\ &\qquad = nu^{n-1}\Phi(x_1, \ldots),\end{aligned}$$

and if we set $u = 1$, then we obtain

$$x_1 \frac{\partial\Phi}{\partial x_1} + x_2 \frac{\partial\Phi}{\partial x_2} + \cdots + x_m \frac{\partial\Phi}{\partial x_m} = n\Phi. \qquad (II)$$

Lecture II (April 29, 1897)

There are two ways in which one can classify forms: either according to order or according to the number of variables—which is more commonly done. In the latter case there are special names for forms with a small number of variables:

1. $m = 1$. *Unary forms.* The general form is cx_1^n.
2. $m = 2$. *Binary forms.* Here the general expression is

$$\mathcal{F} = c_0 x_1^n + c_1 x_1^{n-1} x_2 + \cdots + c_n x_2^n.$$

   The number of terms is $n + 1$.
3. $m = 3$. *Ternary forms.*
4. $m = 4$. *Quaternary forms.*
5. $m = 5$. *Quinary forms.*
6. $m = 6$. *Senary forms.*

The terminology for $m = 1, 5, 6$ is little used however. As the primary purpose of this section we are left with the *determination of the number of terms of a general form.* Let $\phi(n, m)$ be the number of terms of a general form of order n with m variables. We know—and it is easily seen—that

$$\phi(n, 2) = n + 1,$$
$$\phi(1, m) = m.$$

We claim now that the desired formula is

$$\phi(n, m) = \frac{(n+1)(n+2)(n+3)\cdots(n+m-1)}{1 \cdot 2 \cdot 3 \cdots (m-1)}. \tag{III}$$

This formula is valid for $m = 2$. We will prove it for general m by induction from $m - 1$ to m. For this purpose we will use a recursion formula which can be derived as follows. Let $\mathcal{F}^{(n)}$ denote a form of order n; then in the general case we have the expression

$$\mathcal{F}^{(n)}(x_1, \ldots, x_m) = x_m^n \mathcal{F}^{(0)}(x_1, \ldots, x_{m-1}) + x_m^{n-1} \mathcal{F}^{(1)}(x_1, \ldots, x_{m-1}) + \cdots + x_m^0 \mathcal{F}^{(n)}(x_1, \ldots, x_{m-1}).$$

Therefore

$$\phi(n, m) = \phi(0, m-1) + \phi(1, m-1) + \cdots + \phi(n, m-1),$$

and, likewise,

$$\phi(n-1,m) = \phi(0,m-1) + \phi(1,m-1) + \cdots + \phi(n-1,m-1).$$

Subtraction of the two formulas results in

$$\phi(n,m) - \phi(n-1,m) = \phi(n,m-1)$$

or

$$\phi(n,m) = \phi(n-1,m) + \phi(n,m-1).$$

If we now set

$$\chi(n,m) = \phi(n,m) - \frac{(n+1)\cdots(n+m-1)}{1\cdot 2\cdot 3\cdots(m-1)},$$

then we obtain

$$\begin{aligned}\chi(n,m) - \chi(n-1,m) &= \phi(n,m) - \phi(n-1,m)\\ &\quad - \frac{(n+1)\cdots(n+m-1)}{1\cdot 2\cdots(m-1)}\\ &\quad + \frac{n(n+1)\cdots(n+m-2)}{1\cdot 2\cdots(m-1)}\\ &= \phi(n,m-1) - \frac{(n+1)(n+2)\cdots(n+m-2)}{1\cdot 2\cdots(m-2)}\\ &= 0,\end{aligned}$$

since we assume the formula to hold for $m-1$. Consequently,

$$\chi(n,m) = \chi(n-1,m) = \cdots = \chi(1,m),$$

through repeated application of the last formula. Hence, we obtain that

$$\chi(n,m) = \phi(1,m) - \frac{2\cdot 3\cdots m}{2\cdot 3\cdots(m-1)} = m - m = 0.$$

Therefore, formula (III) is valid for m variables, and is thus valid in general.

Lecture III (April 30, 1897)

Finally, we want to point out briefly a geometrical interpretation of the theory of forms. If, given a binary form, we set $x_1 = x$, $x_2 = 1$ (which does not change the essential nature of the form, due to its homogeneity), then we obtain a polynomial in one variable of order n:

$$c_0 x^n + c_1 x^{n-1} + c_2 x^{n-2} + \cdots + c_n.$$

The theory of binary forms therefore includes the theory of algebraic equations in one variable. If we interpret x_1, x_2 as the coordinates of a point on a line, then a binary form set equal to zero represents n points on a line. The theory of binary forms is thus also identical with the geometry of a line (or a bundle of lines and planes). Analogously, one realizes that a ternary form set equal to zero describes a relationship between two variables; hence the theory of ternary forms is identical to the geometry of the plane, namely, of algebraic curves. Finally, the theory of quaternary forms is analogously an essential aid in studying the geometry of space, especially of algebraic surfaces. The forms with more than three variables do not readily admit such a geometric interpretation.

I.2 The linear transformation

We are led to essentially new and deep properties of forms through the application of linear transformations. Let

$$\mathcal{F}^{(n)}(x_1, x_2, \ldots, x_m)$$

be a general form. We can derive another form from it, if we replace the m variables x by other variables x' via relations of the form

$$\begin{aligned} x_1 &= \phi_1(x'_1, \ldots, x'_m), \\ x_2 &= \phi_2(x'_1, \ldots, x'_m), \\ &\cdots \\ x_m &= \phi_m(x'_1, \ldots, x'_m), \end{aligned}$$

where the ϕs denote forms of the same order. Such an operation is called a *transformation*, the new variables are called x', and the resulting form

$$\begin{aligned} &\mathcal{F}'(x'_1, x'_2, \ldots, x'_m) \\ &\quad = \mathcal{F}(\phi_1(x'_1, \ldots, x'_m), \phi_2(x'_1, \ldots, x'_m), \ldots, \phi_m(x'_1, \ldots, x'_m)) \end{aligned}$$

is called the *transformed form.*

Among the transformations, the linear ones are distinguished, that is, the transformations for which $\phi_1, \phi_2, \ldots, \phi_m$ are linear forms. The relationship is then given by the following equations—here again we want to fix notation:

$$x_1 = \alpha_{11}x'_1 + \alpha_{12}x'_2 + \cdots + \alpha_{1m}x'_m,$$
$$x_2 = \alpha_{21}x'_1 + \alpha_{22}x'_2 + \cdots + \alpha_{2m}x'_m,$$
$$\cdots$$
$$x_m = \alpha_{m1}x'_1 + \alpha_{m2}x'_2 + \cdots + \alpha_{mm}x'_m.$$

The α are called the *transformation coefficients.* We regard them as given, without ever specifying them. The transformed form becomes

$$\begin{aligned}\mathcal{F}'(x'_1, x'_2, \ldots, x'_m) &= \mathcal{F}(x_1, x_2, \ldots, x_m)\\ &= \mathcal{F}(\alpha_{11}x'_1 + \cdots, \ldots, \alpha_{m1}x'_1 + \cdots).\end{aligned}$$

Here, we primarily need to observe three properties of linear transformations:

1. The transformed form has the same order as the original form. This is because the general term

$$c_i x_1^{\nu_1} x_2^{\nu_2} \cdots x_m^{\nu_m}, \qquad \nu_1 + \nu_2 + \cdots + \nu_m = n,$$

becomes

$$c_i(\alpha_{11}x'_1 + \alpha_{12}x'_2 + \cdots)^{\nu_1}(\alpha_{21}x'_1 + \cdots)^{\nu_2} \cdots (\alpha_{m1}x'_1 + \cdots)^{\nu_m},$$

from which it is apparent that in each new term the coefficients c_i appear homogeneously to the first power, the transformation coefficients α appear homogeneously to the nth power, and the variables x' also appear homogeneously to the nth power. We now want to write the transformed form exactly like the original one, only with primed letters, namely:

$$\mathcal{F}(x_1, x_2, \ldots, x_m) = c_0 x_1^n + c_1 x_1^{n-1} x_2 + c_2 x_1^{n-2} x_2^2 + c_3 x_1^{n-1} x_3 + \cdots$$

equals

$$\mathcal{F}'(x'_1, x'_2, \ldots, x'_m) = c'_0 {x'_1}^n + c'_1 {x'_1}^{n-1} x'_2 + c'_2 {x'_1}^{n-2} x'^2_2 + c'_3 {x'_1}^{n-1} x'_3 + \cdots .$$

Observe that not only is the stated theorem valid, but we have also shown that the new coefficients c'_i are homogeneous functions of degree one of the original coefficients, and homogeneous functions of degree n of the transformation coefficients α.

A simple example might serve as an illustration.

Let

$$\mathcal{F}^{(2)}(x_1, x_2) = Ax_1^2 + Bx_1x_2 + Cx_2^2.$$

Using the transformation

$$x_1 = \alpha_{11}x_1' + \alpha_{12}x_2',$$
$$x_2 = \alpha_{21}x_1' + \alpha_{22}x_2',$$

we obtain

$$\mathcal{F} = A'x_1'^2 + B'x_1'x_2' + C'x_2'^2 = \mathcal{F}'^{(2)}(x_1', x_2'),$$

where

$$A' = A\alpha_{11}^2 + B\alpha_{11}\alpha_{21} + C\alpha_{21}^2,$$
$$B' = 2A\alpha_{11}\alpha_{12} + B(\alpha_{11}\alpha_{22} + \alpha_{12}\alpha_{21}) + 2C\alpha_{21}\alpha_{22},$$
$$C' = A\alpha_{12}^2 + B\alpha_{12}\alpha_{22} + C\alpha_{22}^2.$$

This confirms the assertions made above.

2. The transformation is invertible. Indeed, to recover the original form from the transformed one, we only need to solve the above equations for the x', which is possible since there are m equations for the m variables x'. We only need to assume—and we want to keep this assumption; it is the only one we need to make—that the *transformation determinant*, that is, the determinant of the transformation coefficients

$$A = \begin{vmatrix} \alpha_{11} & \alpha_{12} & \cdots & \alpha_{1m} \\ \alpha_{21} & \alpha_{22} & \cdots & \alpha_{2m} \\ & & \cdots & \\ \alpha_{m1} & \alpha_{m2} & \cdots & \alpha_{mm} \end{vmatrix},$$

is different from zero. If the $(m-1)\times(m-1)$ minors of this determinant are A_{ik}, then the solved equations are

$$Ax_1' = A_{11}x_1 + A_{21}x_2 + \cdots + A_{m1}x_m,$$
$$Ax_2' = A_{12}x_1 + A_{22}x_2 + \cdots + A_{m2}x_m,$$
$$\cdots$$
$$Ax_m' = A_{1m}x_1 + A_{2m}x_2 + \cdots + A_{mm}x_m.$$

The pattern is very easy to surmise; compared to the other equations the indices are simply transposed. We have

$$x_i = \alpha_{i1}x_1' + \alpha_{i2}x_2' + \cdots + \alpha_{im}x_m',$$
$$Ax_i' = A_{1i}x_1 + A_{2i}x_2 + \cdots + A_{mi}x_m.$$

Lecture IV (May 3, 1897)

3. The group property of linear transformations. If we first apply the transformation

$$x_1 = \alpha_{11}x_1' + \alpha_{12}x_2' + \cdots + \alpha_{1m}x_m',$$
$$\cdots$$
$$x_m = \alpha_{m1}x_1' + \alpha_{m2}x_2' + \cdots + \alpha_{mm}x_m'$$

to a form $\mathcal{F}$, then we obtain the transformed form $\mathcal{F}'^{(n)}(x_1', x_2')$. To this form we now want to apply a second linear transformation, such as

$$x_1' = \beta_{11}x_1'' + \beta_{12}x_2'' + \cdots + \beta_{1m}x_m'',$$
$$\cdots$$
$$x_1' = \beta_{m1}x_1'' + \beta_{m2}x_2'' + \cdots + \beta_{mm}x_m''.$$

In this way we obtain a new form $\mathcal{F}''^{(n)}(x_1'', x_2'')$. The order of this form is the same as that of the original form. It is now clear that one can replace the two steps by a single one, since one has

$$\begin{aligned} x_i = {} & \alpha_{i1}(\beta_{11}x_1'' + \beta_{12}x_2'' + \cdots + \beta_{1m}x_m'') \\ & + \alpha_{i2}(\beta_{21}x_1'' + \beta_{22}x_2'' + \cdots + \beta_{2m}x_m'') \\ & \cdots \\ & + \alpha_{im}(\beta_{m1}x_1'' + \beta_{m2}x_2'' + \cdots + \beta_{mm}x_m''). \end{aligned}$$

From this it is immediately clear that one can obtain the x'' from the x directly through *one* linear transformation, namely, through the following one:

$$x_1 = \gamma_{11}x_1'' + \gamma_{12}x_2'' + \cdots + \gamma_{1m}x_m'',$$
$$\cdots$$
$$x_m = \gamma_{m1}x_1'' + \gamma_{m2}x_2'' + \cdots + \gamma_{mm}x_m'',$$

where we set

$$\gamma_{il} = \alpha_{i1}\beta_{1l} + \alpha_{i2}\beta_{2l} + \cdots + \alpha_{ik}\beta_{kl} + \cdots + \alpha_{im}\beta_{ml}\,.$$

Consequently, one also obtains:

$$\begin{vmatrix} \alpha_{11} & \alpha_{12} & \cdots & \alpha_{1m} \\ \alpha_{21} & \alpha_{22} & \cdots & \alpha_{2m} \\ & & \cdots & \\ \alpha_{m1} & \alpha_{m2} & \cdots & \alpha_{mm} \end{vmatrix} \cdot \begin{vmatrix} \beta_{11} & \beta_{12} & \cdots & \beta_{1m} \\ \beta_{21} & \beta_{22} & \cdots & \beta_{2m} \\ & & \cdots & \\ \beta_{m1} & \beta_{m2} & \cdots & \beta_{mm} \end{vmatrix}$$

$$= \begin{vmatrix} \gamma_{11} & \gamma_{12} & \cdots & \gamma_{1m} \\ \gamma_{21} & \gamma_{22} & \cdots & \gamma_{2m} \\ & & \cdots & \\ \gamma_{m1} & \gamma_{m2} & \cdots & \gamma_{mm} \end{vmatrix},$$

which, incidentally, can also be concluded from the simultaneous vanishing of A (respectively, B and C).

From the above we obtain the following theorem, which we call the *group property of linear transformations*:

Theorem *A sequence of linear transformations can be replaced by a single linear transformation.*

In the case of two linear transformations, each new coefficient γ is a linear homogeneous function of the α, as well as the β.

Geometrically, a linear transformation means a projective relationship. The equation $\mathcal{F}^{(n)}(x_1, \ldots, x_m) = 0$, for $m = 2, 3, 4$, represents a point set, a curve, or a surface of order n, respectively; $\mathcal{F}'^{(n)}(x_1, \ldots, x_m)$ represents a similar object, related to the previous one by colinearity. The first property, therefore, says that the order of two objects that are related through colinearity is equal; the second one says that the colinear relationship is mutual; the third says that if two objects are related by colinearity to a third one, then they are related colinearly to each other.

If we set

$$\mathcal{F}^{(n)}(x_1, \ldots, x_m) = a_1 x_1^n + \cdots,$$

$$\mathcal{F}'^{(n)}(x'_1, \ldots, x'_m) = a'_1 x'^n_1 + \cdots,$$

$$\mathcal{F}''^{(n)}(x''_1, \ldots, x''_m) = a''_1 x''^n_1 + \cdots,$$

then the a''_i are, as remarked above (Lecture III), linear in the a'_i, of nth order in the β_{ik}, and hence also homogeneous linear in the a_i and homogeneous of nth order in the α as well as the β.

The example $n = 2, m = 2$ used above shall again serve as an illustration. There (Lecture III) we had

$$\mathcal{F}^{(2)}(x_1, x_2) = Ax_1^2 + Bx_1x_2 + Cx_2^2$$

transformed into

$$\mathcal{F} = \mathcal{F}' = A'x_1'^2 + B'x_1'x_2' + C'x_2'^2$$

via the transformation

$$\begin{aligned} x_1 &= \alpha_{11}x_1' + \alpha_{12}x_2', \\ x_2 &= \alpha_{21}x_1' + \alpha_{22}x_2', \end{aligned}$$

where

$$\begin{aligned} A' &= A\alpha_{11}^2 + B\alpha_{11}\alpha_{21} + C\alpha_{21}^2, \\ B' &= 2A\alpha_{11}\alpha_{12} + B(\alpha_{11}\alpha_{22} + \alpha_{12}\alpha_{21}) + 2C\alpha_{21}\alpha_{22}, \\ C' &= A\alpha_{12}^2 + B\alpha_{12}\alpha_{22} + C\alpha_{22}^2. \end{aligned}$$

Therefore, the transformation

$$\begin{aligned} x_1' &= \beta_{11}x_1'' + \beta_{12}x_2'', \\ x_2' &= \beta_{21}x_1'' + \beta_{22}x_2'' \end{aligned}$$

applied to the function $\mathcal{F}'$ results in

$$\mathcal{F} = \mathcal{F}' = \mathcal{F}'' = A''x_1''^2 + B''x_1''x_2'' + C''x_2''^2,$$

where

$$\begin{aligned} A'' &= A'\beta_{11}^2 + B'\beta_{11}\beta_{21} + C'\beta_{21}^2, \\ B'' &= 2A'\beta_{11}\beta_{12} + B'(\beta_{11}\beta_{22} + \beta_{12}\beta_{21}) + 2C'\beta_{21}\beta_{22}, \\ C'' &= A'\beta_{12}^2 + B'\beta_{12}\beta_{22} + C'\beta_{22}^2, \end{aligned}$$

or, if we use the expansions for A', B', C':

$$\begin{aligned} A'' = A(&\alpha_{11}^2\beta_{11}^2 + 2\alpha_{11}\alpha_{12}\beta_{11}\beta_{21} + \alpha_{12}^2\beta_{21}^2) \\ &+ B(\alpha_{11}\alpha_{21}\beta_{11}^2 + (\alpha_{11}\alpha_{22} + \alpha_{12}\alpha_{21})\beta_{11}\beta_{21} + \alpha_{12}\alpha_{22}\beta_{21}^2) \\ &+ C(\alpha_{21}^2\beta_{11}^2 + 2\alpha_{11}\alpha_{22}\beta_{11}\beta_{21} + \alpha_{22}^2\beta_{21}^2), \end{aligned}$$

$$\begin{aligned}B'' = {} & 2A(\alpha_{11}^2\beta_{11}\beta_{12} + \alpha_{11}\alpha_{12}(\beta_{11}\beta_{22} + \beta_{21}\beta_{12}) + \alpha_{12}^2\beta_{21}\beta_{22}) \\ & + B(2\alpha_{11}\alpha_{21}\beta_{11}\beta_{12} + (\alpha_{11}\alpha_{22} + \alpha_{12}\alpha_{21})(\beta_{11}\beta_{22} + \beta_{12}\beta_{21}) \\ & \quad + 2\alpha_{12}\alpha_{22}\beta_{21}\beta_{22}) \\ & + 2C(\alpha_{21}^2\beta_{11}\beta_{12} + \alpha_{21}\alpha_{22}(\beta_{11}\beta_{22} + \beta_{21}\beta_{12}) + \alpha_{22}^2\beta_{21}\beta_{22}), \\ C'' = {} & A(\alpha_{11}^2\beta_{12}^2 + 2\alpha_{11}\alpha_{12}\beta_{12}\beta_{22} + \alpha_{12}^2\beta_{22}^2) \\ & + B(\alpha_{11}\alpha_{21}\beta_{12}^2 + (\alpha_{11}\alpha_{22} + \alpha_{12}\alpha_{21})\beta_{12}\beta_{22} + \alpha_{12}\alpha_{22}\beta_{22}^2) \\ & + C(\alpha_{21}^2\beta_{12}^2 + 2\alpha_{21}\alpha_{22}\beta_{12}\beta_{22} + \alpha_{22}^2\beta_{22}^2).\end{aligned}$$

One can also arrive at these formulas directly by setting

$$\begin{aligned}A'' &= A\gamma_{11}^2 + B\gamma_{11}\gamma_{21} + C\gamma_{21}^2, \\ B'' &= 2A\gamma_{11}\gamma_{12} + B(\gamma_{11}\gamma_{22} + \gamma_{12}\gamma_{21}) + 2C\gamma_{21}\gamma_{22}, \\ C'' &= A\gamma_{12}^2 + B\gamma_{12}\gamma_{22} + C\gamma_{22}^2,\end{aligned}$$

where

$$\begin{aligned}\gamma_{11} &= \alpha_{11}\beta_{11} + \alpha_{12}\beta_{21}, \\ \gamma_{12} &= \alpha_{11}\beta_{12} + \alpha_{12}\beta_{22}, \\ \gamma_{21} &= \alpha_{21}\beta_{11} + \alpha_{22}\beta_{21}, \\ \gamma_{22} &= \alpha_{21}\beta_{12} + \alpha_{22}\beta_{22}.\end{aligned}$$

Indeed, the combination of the two transformations results in

$$\begin{aligned}x_1 &= \alpha_{11}(\beta_{11}x_1'' + \beta_{12}x_2'') + \alpha_{12}(\beta_{21}x_1'' + \beta_{22}x_2'') = \gamma_{11}x_1'' + \gamma_{12}x_2'', \\ x_2 &= \alpha_{21}(\beta_{11}x_1'' + \beta_{12}x_2'') + \alpha_{22}(\beta_{21}x_1'' + \beta_{22}x_2'') = \gamma_{21}x_1'' + \gamma_{22}x_2''.\end{aligned}$$

If one substitutes the formulas for γ into the expressions preceding them, then one indeed again obtains the above formulas. One may also verify the remarks regarding the orders.

Lecture V (May 4, 1897)

Since a sequence of linear transformations may be replaced by a single linear transformation, the question arises whether, conversely, a general linear transformation may be replaced by a sequence of special linear transformations. To begin with, it is clear that this is possible in different ways; one has to make appropriate choices for the special transformations. Here we only treat the case of binary forms since this is the most useful for us, and the general case is not fundamentally different.

We make the following claim:

Claim *Every linear transformation of binary forms can be composed of the following three types of linear transformations:*

$$(1)\quad x_1 = \kappa x_1', \qquad x_2 = \lambda x_2'.$$

$$(2)\quad x_1 = x_1' + \mu x_2', \qquad x_2 = x_2'.$$

$$(3)\quad x_1 = x_1', \qquad x_2 = \nu x_1' + x_2'.$$

Aside from these, we mention the important linear transformation

$$(3')\quad x_1 = x_2', \qquad x_2 = x_1'.$$

Our goal is to prove the asserted theorem, namely, to obtain from (1), (2), (3) the general linear transformation

$$x_1 = \alpha_{11} x_1' + \alpha_{12} x_2',$$
$$x_2 = \alpha_{21} x_1' + \alpha_{22} x_2'.$$

If we apply the transformations (1), (2), (3) successively, then we obtain, by omitting the primes in the intermediate steps after applying a transformation:

$$x_1 = \kappa x_1' \qquad \kappa x_1 = \kappa(x_1' + \mu x_2')$$
$$x_2 = \lambda x_2' \qquad \lambda x_2 = \lambda x_2'$$

$$\kappa(x_1 + \mu x_2) = \kappa(x_1' + \mu\nu x_1' + \mu x_2'),$$
$$\lambda x_2 = \lambda\nu x_1' + \lambda x_2'$$

or

$$x_1 = (\kappa + \kappa\mu\nu) x_1' + \kappa\mu x_2',$$
$$x_2 = \lambda\nu x_1' + \lambda x_2'.$$

This can be viewed as the most general linear transformation, because one only needs to set

$$\lambda = \alpha_{22},$$
$$\nu = \frac{\alpha_{21}}{\alpha_{22}},$$
$$\mu = \frac{\alpha_{12}\alpha_{22}}{\alpha_{11}\alpha_{22} - \alpha_{12}\alpha_{21}},$$
$$\kappa = \frac{\alpha_{11}\alpha_{22} - \alpha_{12}\alpha_{21}}{\alpha_{22}}$$

in order to obtain the initial form. However, if one of the denominators becomes zero, then this method does not work. If $\alpha_{11}\alpha_{22} - \alpha_{12}\alpha_{21} = 0$, then one has

$$\delta = \begin{vmatrix} \alpha_{11} & \alpha_{12} \\ \alpha_{21} & \alpha_{22} \end{vmatrix} = 0,$$

that is, the transformation determinant vanishes, while we explicitly made the assumption (Lecture III) that it should not vanish.

On the other hand, the case

$$\alpha_{22} = 0$$

needs to be considered separately; then we must assume that, because $\delta \neq 0$,

$$\alpha_{12} \neq 0, \qquad \alpha_{21} \neq 0.$$

To obtain the special transformation

$$x_1 = \alpha_{11}x'_1 + \alpha_{12}x'_2,$$
$$x_2 = \alpha_{21}x'_1$$

via the transformations (1), (2), (3), we first derive (3′) from (1), (2), (3), which is done through successive application of

(3) for $\nu = 1$; (2) for $\mu = -1$; (3) for $\nu = 1$; (1) for $\kappa = +1$, $\lambda = -1$.

Because then one obtains

$$x_1 = x'_1 \qquad\qquad x_1 = x'_1 - x'_2$$
$$x_2 = x'_1 + x'_2 \qquad\qquad x_1 + x_2 = x'_1 - x'_2 + x'_2$$

$$x_1 - x_2 = x'_1 - x'_1 - x'_2 \qquad\qquad -x_2 = x'_2$$
$$x_1 = x'_1 \qquad\qquad x_1 = x'_1$$

that is,

$$x_1 = x_2',$$
$$x_2 = x_1'.$$

This is the transformation (3′). According to the preceding, one can now derive from (1), (2), (3) the transformation

$$x_1 = \alpha_{12}x_1' + \alpha_{11}x_2',$$
$$x_2 = \alpha_{21}x_2',$$

since $\alpha_{12} \neq 0$, $\alpha_{21} \neq 0$. But transformation (3′) takes this transformation to

$$x_1 = \alpha_{11}x_1' + \alpha_{12}x_2',$$
$$x_2 = \alpha_{21}x_1',$$

which completely proves the theorem above.

On the other hand, one can obtain the transformation (3) from (2), (3′) through application of the transformations (3′), (2) for $\mu = \nu$, (3′). This is because we obtain

$$\begin{array}{lll} x_1 = x_2' & x_2 = x_2' & x_2 = x_1', \\ x_2 = x_1' & x_1 = x_1' + \nu x_2' & x_1 + \nu x_2 = x_2' + \nu x_1' \end{array}$$

and so indeed

$$x_1 = x_1',$$
$$x_2 = \nu x_1' + x_2'.$$

And this, in turn, implies the following theorem.

Theorem *Every linear transformation can be obtained through the transformations* (1), (2), (3′).

If one has two given numerical substitutions, one can, through the combination of both, derive arbitrarily many other transformations (even through repeated application of a single substitution). Nevertheless, in general one will not obtain all possible transformations in this way. Instead, one will obtain a certain *group of transformations* through such mutual combinations. In general, the number of transformations in this group will be infinite; only in special cases will it be finite. In the case of binary forms, these finite transformation groups are intimately connected with the regular solids. We cannot go into details at this time, however.

I.3 The concept of an invariant

In the following we will initially limit ourselves to the theory of binary forms and use them to clarify the general concepts. The generalization of the latter to forms with arbitrarily many variables poses no difficulties in most cases.

We always want to write the general binary form as:

$$f^{(n)}(x_1, x_2) = a_0 x_1^n + \binom{n}{1} a_1 x_1^{n-1} x_2 + \binom{n}{2} a_2 x_1^{n-2} x_2^2 + \cdots + a_n x_2^n,$$

where $\binom{n}{i}$ are the binomial coefficients

$$\binom{n}{i} = \frac{n(n-1)(n-2)\cdots(n-i+1)}{1 \cdot 2 \cdot 3 \cdots i}.$$

We will always use the word "coefficients" for the a_i in this form (not multiplied by the binomial coefficients).

Application of the linear transformation

$$\begin{aligned} x_1 &= \alpha_{11} x_1' + \alpha_{12} x_2', \\ x_2 &= \alpha_{21} x_1' + \alpha_{22} x_2' \end{aligned}$$

produces a new form $f'^{(n)}(x_1', x_2')$. Similarly, application of the linear transformation

$$\begin{aligned} x_1 &= \beta_{11} x_1'' + \beta_{12} x_2'', \\ x_2 &= \beta_{21} x_1'' + \beta_{22} x_2'' \end{aligned}$$

produces another form $f''^{(n)}(x_1'', x_2'')$, etc. The question now arises: Do there exist properties that are common to all these "*equivalent*" forms, which are derived from f through all possible linear transformations?

Such properties produce the invariants and covariants of the form f, which we now have to define and inspect more closely.

Definition *An invariant of the base form f is a polynomial function of the coefficients $a_0, a_1, \ldots, a_n$ that changes only by a factor equal to a power of the transformation determinant δ if one replaces the coefficients $a_0, a_1, \ldots, a_n$ of the given base form by the corresponding coefficients $a_0', a_1', \ldots, a_n'$ of the linearly transformed form.*

Lecture VI (May 6, 1897)

The defining equation of the invariants is therefore

$$\mathcal{I}(a_0', a_1', \ldots, a_n') = \delta^p \mathcal{I}(a_0, a_1, \ldots, a_n).$$

Here we assume the base form to be given as

$$f = a_0 x_1^n + \binom{n}{1} a_1 x_1^{n-1} x_2 + \cdots + a_n x_2^n,$$

and the transformed form as

$$f' = a_0' {x_1'}^n + \binom{n}{1} a_1' {x_1'}^{n-1} x_2' + \cdots + a_n' {x_2'}^n.$$

Furthermore, the transformation

$$\begin{aligned} x_1 &= \alpha_{11} x_1' + \alpha_{12} x_2', \\ x_2 &= \alpha_{21} x_1' + \alpha_{22} x_2' \end{aligned}$$

has $\delta = \alpha_{11}\alpha_{22} - \alpha_{12}\alpha_{21}$.

The quadratic form

$$f = a_0 x_1^2 + 2a_1 x_1 x_2 + a_2 x_2^2$$

provides an example. The transformed form is:

$$f' = a_0' {x_1'}^2 + 2a_1' x_1' x_2' + a_2' {x_2'}^2,$$

where (cf. Lecture III):

$$\begin{aligned} a_0' &= a_0 \alpha_{11}^2 + 2a_1 \alpha_{11}\alpha_{21} + a_2 \alpha_{21}^2, \\ a_1' &= a_0 \alpha_{11}\alpha_{12} + a_1(\alpha_{11}\alpha_{22} + \alpha_{12}\alpha_{21}) + a_2 \alpha_{21}\alpha_{22}, \\ a_2' &= a_0 \alpha_{12}^2 + 2a_1 \alpha_{12}\alpha_{22} + a_2 \alpha_{22}^2. \end{aligned}$$

One invariant of the form f is the expression

$$a_0 a_2 - a_1^2,$$

because one has

$$\begin{aligned} & a_0' a_2' - {a_1'}^2 \\ &= a_0^2(\alpha_{11}^2 \alpha_{12}^2 - \alpha_{11}^2 \alpha_{12}^2) \\ &\quad + 2a_0 a_1 \big(\alpha_{11}^2 \alpha_{12}\alpha_{22} + \alpha_{12}^2 \alpha_{11}\alpha_{21} - \alpha_{11}\alpha_{12}(\alpha_{11}\alpha_{22} + \alpha_{12}\alpha_{21})\big) \\ &\quad + a_0 a_2 (\alpha_{11}^2 \alpha_{22}^2 + \alpha_{12}^2 \alpha_{21}^2 - 2\alpha_{11}\alpha_{12}\alpha_{21}\alpha_{22}) \\ &\quad + a_1^2 \big(4\alpha_{11}\alpha_{21}\alpha_{12}\alpha_{22} - (\alpha_{11}\alpha_{22} + \alpha_{12}\alpha_{21})^2\big) \\ &\quad + a_2^2 (\alpha_{21}^2 \alpha_{22}^2 - \alpha_{21}^2 \alpha_{22}^2) \\ &\quad + 2a_1 a_2 \big(\alpha_{11}\alpha_{21}\alpha_{22}^2 + \alpha_{12}\alpha_{22}\alpha_{21}^2 - \alpha_{21}\alpha_{22}(\alpha_{11}\alpha_{22} + \alpha_{12}\alpha_{21})\big) \\ &= a_0 a_2 (\alpha_{11}\alpha_{22} - \alpha_{12}\alpha_{21})^2 - a_1^2 (\alpha_{11}\alpha_{22} - \alpha_{12}\alpha_{21})^2 \\ &= \delta^2 (a_0 a_2 - a_1^2), \end{aligned}$$

as asserted.

An invariant set equal to zero defines a property that is shared by the whole class of equivalent forms. This is of essential importance. In our example,

$$a_0a_2 - a_1^2 = 0$$

indicates that the equation $f = 0$ in x_1/x_2 has a double root or, what amounts to the same thing, that f is the square of a linear form, a property that is indeed preserved under linear transformation. Because, if

$$f = (l_0x_1 + l_1x_2)^2,$$

then

$$\begin{aligned} f' &= \bigl(l_0(\alpha_{11}x_1' + \alpha_{12}x_2') + l_1(\alpha_{21}x_1' + \alpha_{22}x_2')\bigr)^2 \\ &= (l_0'x_1' + l_1'x_2')^2. \end{aligned}$$

In addition to the invariants we need to consider the covariants, which we define as follows:

Definition *A covariant is a polynomial function of the coefficients* a_0, $a_1, \ldots, a_n$ *and the variables* x_1, x_2 *that changes only by a factor equal to a power of the transformation determinant* δ *if one replaces the coefficients* $a_0, a_1, \ldots, a_n$ *of the given base form by the corresponding coefficients* $a_0', a_1', \ldots, a_n'$ *of the linearly transformed form and, simultaneously, replaces the variables* x_1, x_2 *by the linearly transformed variables* x_1', x_2'.

The defining equation of the covariant is therefore

$$C(a_0', a_1', \ldots, a_n'; x_1', x_2') = \delta^p C(a_0, a_1, \ldots, a_n; x_1, x_2).$$

The power with which the coefficients $a_0, a_1, \ldots, a_n$ appear will always be called the *degree* of the invariant; the power of the variables x_1, x_2 will always be called the *order* of the covariant or invariant. An invariant is then simply a covariant whose order is zero. Invariants and covariants together are also called an "invariant system," and their characterizing property the "invariant property."

The simplest covariant is the form f itself. In this case $p = 0$.

One can see in the example of an invariant we gave that it is very difficult to recognize the invariant property of an invariant or covariant by means of calculation. We therefore have to look now for properties of invariants and covariants until we have found those that characterize them precisely. Then we will succeed in proving the invariant property of a given function, and, moreover, to list all possible in- and covariants.

Lecture VII (May 7, 1897)

I.4 Properties of invariants and covariants

Before we consider special properties of in- and covariants, we remark that, without loss of generality, they may be assumed to be homogeneous in the x as well as the a. This is because, according to Section I.2, after a linear transformation is applied, each x is replaced by a linear combination x' of the x, and likewise $a_0, a_1, \ldots, a_n$ are replaced by linear combinations of these coefficients. Therefore, each homogeneous polynomial function of the x on the one hand, and of $a_0, a_1, \ldots, a_n$ on the other hand, remains homogeneous after a linear transformation is applied. Therefore, even if $C(a_0, a_1, \ldots, a_n; x_1, x_2)$ is not a homogeneous function of the a and the x, then the invariant property also holds for its homogeneous parts, since the equation $C(a'_0, a'_1, \ldots, a'_n; x'_1, x'_2) = \delta^p C(a_0, a_1, \ldots, a_n; x_1, x_2)$ has to hold for each such part of C as well. Therefore, we can and will, in the following, assume each in- and covariant to be homogeneous in x as well as a. We will show, however, that each invariant of a form always has to be homogeneous in the a to begin with; furthermore, that each covariant that is assumed to be homogeneous in the x necessarily has to be homogeneous in the a.

In this section we shall pursue the following strategy for the derivation of the invariant properties of in- and covariants. A function that possesses the invariant property must also possess it for each of the three types of transformations. Through application of each of the latter we will arrive at important relations. But since a general linear transformation is composed of these three types, application of a general transformation will in itself not lead to any new results. On the other hand, one consequence of this fact will be that in- and covariants are even characterized by these properties. Thus, we will successively apply the three types.

Apply the transformation (1) (Lecture IV) to the form

$$f(x_1, x_2) = a_0 x_1^n + \binom{n}{1} a_1 x_1^{n-1} x_2 + \cdots + \binom{n}{i} a_i x_1^{n-i} x_2^i + \cdots + a_n x_2^n.$$

The given form then becomes:

$$f = f(\kappa x_1', \lambda x_2') = a_0\kappa^n {x_1'}^n + \binom{n}{1} a_1 \kappa^{n-1} {x_1'}^{n-1} \lambda x_2' + \cdots$$
$$+ \binom{n}{i} a_i \kappa^{n-i} {x_1'}^{n-i} \lambda^i {x_2'}^i + \cdots + a_n \lambda^n {x_2'}^n$$
$$= a_0' {x_1'}^n + \binom{n}{1} a_1' {x_1'}^{n-1} x_2' + \cdots + \binom{n}{i} a_i' {x_1'}^{n-i} {x_2'}^i$$
$$+ \cdots + a_n' {x_2'}^n,$$

where

$$\begin{aligned} a_0' &= a_0\kappa^n, \\ a_1' &= a_1\kappa^{n-1}\lambda, \\ &\cdots \\ a_i' &= a_i \kappa^{n-i}\lambda^i, \\ &\cdots \\ a_n' &= a_n\lambda^n. \end{aligned}$$

Now, let

$$\mathcal{I}(a_0, a_1, \ldots, a_n) = \sum Z_{\nu_0\nu_1\ldots\nu_n} a_0^{\nu_0} a_1^{\nu_1} \cdots a_n^{\nu_n}$$

be an invariant of the form f, where the Z are numerical coefficients. Then we have the identity

$$\begin{aligned} \mathcal{I}(a_0', a_1', \ldots, a_n') &= \mathcal{I}(a_0\kappa^n, a_1\kappa^{n-1}\lambda, \ldots, a_n\lambda^n) \\ &= \sum \{ Z_{\nu_0\nu_1\ldots\nu_n} a_0^{\nu_0} \kappa^{n\nu_0} a_1^{\nu_1} \kappa^{(n-1)\nu_1} \lambda^{\nu_1} \\ &\quad \cdots a_i^{\nu_i} \kappa^{(n-i)\nu_i} \lambda^{i\nu_i} \cdots a_n^{\nu_n} \lambda^{n\nu_n} \} \\ &= \sum \{ Z_{\nu_0\nu_1\ldots\nu_n} a_0^{\nu_0} a_1^{\nu_1} \cdots a_i^{\nu_i} \cdots a_n^{\nu_n} \\ &\quad \kappa^{n\nu_0 + (n-1)\nu_1 + \cdots + (n-i)\nu_i + \cdots + \nu_{n-1}} \lambda^{\nu_1 + \cdots + i\nu_i + \cdots + n\nu_n} \} \\ &= \kappa^p \lambda^p \sum Z_{\nu_0\nu_1\ldots\nu_n} a_0^{\nu_0} a_1^{\nu_1} \cdots a_n^{\nu_n}, \end{aligned}$$

from which we obtain the identities

$$\begin{aligned} n\nu_0 + (n-1)\nu_1 + \cdots + (n-i)\nu_i + \cdots + \nu_{n-1} &= p, \\ \nu_1 + \cdots + i\nu_i + \cdots + (n-1)\nu_{n-1} + n\nu_n &= p. \end{aligned}$$

Addition of these two formulas results in:

$$n\nu_0 + n\nu_1 + \cdots + n\nu_i + \cdots + n\nu_{n-1} + n\nu_n = 2p,$$

or

$$n(\nu_0 + \nu_1 + \cdots + \nu_n) = 2p.$$

But

$$g = \nu_0 + \nu_1 + \cdots + \nu_n$$

is just the degree of the term under consideration (Lecture VI). We want to call the expression

$$\nu_1 + 2\nu_2 + 3\nu_3 + \cdots + n\nu_n$$

the *weight* of the term under consideration.

The two characteristic equations we have found are the following:

$$\nu_1 + 2\nu_2 + 3\nu_3 + \cdots + n\nu_n = p, \qquad (I)$$

$$ng = 2p. \qquad (II)$$

But since p has to be the same for all terms, we have the following:

Theorem *Every invariant of a form must be homogeneous in the coefficents, and every term must have degree*

$$g = \frac{2p}{n},$$

where p is that exponent of the transformation determinant by which the invariant changes under substitution of the transformed coefficients. Furthermore, all terms must have the same weight, which is also equal to p.

The common weight of all the terms is also called the *weight of the invariant*; thus, this weight gives, at the same time, the exponent of the transformation determinant.

On the other hand, the theorem just proven has evidently a converse. Namely, we have the following:

Theorem *Every homogeneous and isobaric function of the coefficients a, which satisfies the equation $ng - 2p = 0$, where g is the degree, p the weight of this function, and n the order of the base form, is an invariant with respect to transformation (1).**

The proof is evident; if one substitutes the a' instead of the a into $\mathcal{I}$, then a factor $\kappa^p\lambda^p$ appears because of the assumed identities.

* A function is isobaric if all terms have the same weight.

One can verify the theorems in the example of the invariant of a quadratic form:

$$a_0'a_2' - {a_1'}^2 = \delta^2(a_0a_2 - a_1^2).$$

Here, $n = 2, p = 2, g = 2$, and indeed

$$ng = 2 \cdot 2 = 2 \cdot 2 = 2p,$$

and

$$\nu_1 + 2\nu_2 = \nu_1' + 2\nu_2' = 2.$$

We proceed in the exact same fashion to obtain the analogous relations for covariants. We shall assume the covariant

$$\mathcal{C}(a_0, a_1, \ldots, a_n; x_1, x_2)$$

to be homogeneous in the x (see earlier in this Lecture). Let the order of the covariant be m, so that it has the form

$$\mathcal{C}(a_0, a_1, \ldots, a_n; x_1, x_2) = C_0x_1^m + \binom{m}{1}C_1x_1^{m-1}x_2 + \cdots + C_mx_2^m.$$

The coefficients of $\mathcal{C}$ are, as before,

$$C_i = \sum Z_{\nu_0\nu_1\ldots\nu_n}a_0^{\nu_0}a_1^{\nu_1}\cdots a_n^{\nu_n},$$

where the Z are again numbers. From the definition of a covariant we obtain the relation

$$\mathcal{C}(a_0', a_1', \ldots, a_n'; x_1', x_2') = \delta^p\mathcal{C}(a_0, a_1, \ldots, a_n; x_1, x_2),$$

and in particular also

$$\mathcal{C}\left(a_0\kappa^n, a_1\kappa^{n-1}\lambda, \ldots, a_n\lambda^n; \frac{x_1}{\kappa}, \frac{x_2}{\lambda}\right) = \kappa^p\lambda^p\mathcal{C}(a_0, a_1, \ldots, a_n; x_1, x_2).$$

Comparing the coefficients of $x_1^{m-i}x_2^i$ on both sides, one obtains the identity:

$$\sum Z_{\nu_0\nu_1\ldots\nu_n}a_0^{\nu_0}a_1^{\nu_1}\cdots a_n^{\nu_n}\kappa^{n\nu_0+(n-1)\nu_1+\cdots+\nu_{n-1}}$$
$$\times\lambda^{\nu_1+2\nu_2+\cdots+n\nu_n}\kappa^{-(m-i)}\lambda^{-i}$$
$$= \sum \kappa^p\lambda^pZ_{\nu_0\nu_1\ldots\nu_n}a_0^{\nu_0}a_1^{\nu_1}\cdots a_n^{\nu_n}.$$

Thus, we obtain the relations

$$n\nu_0 + (n-1)\nu_1 + \cdots + \nu_{n-1} - (m-i) = p,$$
$$\nu_1 + 2\nu_2 + \cdots + n\nu_n - i = p,$$

and, through addition of these two, the third one

$$n\nu_0 + n\nu_1 + \cdots + n\nu_n - m = 2p,$$

or

$$ng_i - m = 2p,$$

where g_i is the degree of the appropriate term. Therefore, the characteristic equations of a covariant are

$$\nu_1 + 2\nu_2 + \cdots + n\nu_n = p + i, \qquad (III)$$

$$ng_i - 2p = m. \qquad (IV)$$

We can again omit the subscript i in g_i, since the last relation shows that g is constant for all terms; hence it must be equal to the degree of the covariant.

Lecture VIII (May 10, 1897)

Thus, we have the following:

Theorem *If*

$$C = C_0 x_1^m + \binom{m}{1} C_1 x_1^{m-1} x_2 + \binom{m}{2} C_2 x_1^{m-2} x_2^2 + \cdots + C_m x_2^m$$

is a covariant of a form, then all terms of the coefficients $C_0, C_1, \ldots, C_m$ *have the same degree* g*; however, they have weights equal to* $p, p+1, \ldots, p+m$*, respectively. The weight* p *of the first coefficient* C_0 *is called the* **weight of the covariant**, *and determines at the same time the power of the transformation determinant by which the covariant changes under the transformation. The order of the covariant is*

$$m = ng - 2p.$$

For $m = 0$ we obtain from this the theorems about the invariants. Conversely, one also has the following:

Theorem *Each function that is homogeneous in the coefficients on the one hand and the variables on the other hand, and which, when ordered as above, satisfies the equations*

$$\nu_1 + 2\nu_2 + \cdots + n\nu_n = p + i,$$

$$ng - 2p = m,$$

is a covariant with respect to (1).

Because, if we substitute into C the a' and x' instead of the a and x, then a factor of $\kappa^p\lambda^p$ appears, due to the assumed relations.

Analogously, we now have to study which properties we obtain from the application of transformation (2) (Lecture V):

$$\begin{aligned} x_1 &= x_1' + \mu x_2', \\ x_2 &= x_2'. \end{aligned} \qquad (\delta = 1) \tag{2}$$

If we apply it to the form

$$f = a_0 x_1^n + \binom{n}{1} a_1 x_1^{n-1} x_2 + \cdots + a_n x_2^n,$$

then we obtain:

$$\begin{aligned} f' &= a_0(x_1' + \mu x_2')^n + \binom{n}{1} a_1 (x_1' + \mu x_2')^{n-1} x_2' + \cdots \\ &\quad + \binom{n}{i} a_i (x_1' + \mu x_2')^{n-i} {x_2'}^i + \cdots + a_n {x_2'}^n \\ &= a_0' {x_1'}^n + \binom{n}{1} a_1' {x_1'}^{n-1} x_2' + \cdots + \binom{n}{i} a_i' {x_1'}^{n-i} {x_2'}^i + \cdots + a_n' {x_2'}^n, \end{aligned}$$

where

$$\begin{aligned} a_0' &= a_0, \\ a_1' &= a_1 + \mu a_0, \\ a_2' &= a_2 + 2\mu a_1 + \mu^2 a_0, \\ &\cdots. \end{aligned}$$

In general, $a_i' = a_i + \binom{i}{1}\mu a_{i-1} + \binom{i}{2}\mu^2 a_{i-2} + \cdots + \mu^i a_0$. This is because a comparison of the coefficients of ${x_1'}^{n-i}{x_2'}^i$ gives

$$\binom{n}{i} a_i' = \binom{n}{i} a_i + \cdots + \binom{n}{i-k}\binom{n-i+k}{k} a_{i-k}\mu^k + \cdots + \binom{n}{i} a_0 \mu^i.$$

Thus, the above formula follows because of the identity

$$\binom{n}{i-k}\binom{n-i+k}{k} = \binom{n}{i}\binom{i}{k}.$$

Since $\delta = 1$, an invariant has to satisfy the equation

$$\mathcal{I}(a_0, a_1, \ldots, a_n) = \mathcal{I}(a_0', a_1', \ldots, a_n'),$$

where we assume that the above expressions are substituted into the right-hand side. This equation is supposed to hold for all μ, thus one can differentiate both sides with respect to μ and obtains in this way:

$$0 = \frac{\partial \mathcal{I}(a')}{\partial a_0'} \frac{da_0'}{d\mu} + \frac{\partial \mathcal{I}(a')}{\partial a_1'} \frac{da_1'}{d\mu} + \cdots + \frac{\partial \mathcal{I}(a')}{\partial a_n'} \frac{da_n'}{d\mu}.$$

But we have

$$\begin{aligned} \frac{da_0'}{d\mu} &= 0, \\ \frac{da_1'}{d\mu} &= a_0', \\ \frac{da_2'}{d\mu} &= 2a_1', \\ &\cdots, \end{aligned}$$

and, in general, $da_i'/d\mu = ia_{i-1}'$. The function $\mathcal{I}(a_0', a_1', \ldots, a_n')$ therefore satisfies the differential equation

$$\frac{\partial \mathcal{I}(a')}{\partial a_0'} \cdot 0 + \frac{\partial \mathcal{I}(a')}{\partial a_1'} a_0' + \cdots + \frac{\partial \mathcal{I}(a')}{\partial a_n'} na_{n-1}' = 0.$$

But the function $\mathcal{I}(a')$ depends on the a' in the exact same way as $\mathcal{I}(a)$ depends on the a, which gives the following:

Theorem *An invariant of a form satisfies the differential equation*

$$a_0 \frac{\partial \mathcal{I}}{\partial a_1} + 2a_1 \frac{\partial \mathcal{I}}{\partial a_2} + 3a_2 \frac{\partial \mathcal{I}}{\partial a_3} + \cdots + na_{n-1} \frac{\partial \mathcal{I}}{\partial a_n} = 0. \qquad (V)$$

On the other hand, one can also show easily that *each function, which depends only on the coefficients and satisfies the differential equation* (V), *is an invariant with respect to type (2).*

Because, if the differential equation (V) is satisfied, then the equation

$$\frac{\partial \mathcal{I}(a')}{\partial a_1'} a_0' + \cdots + \frac{\partial \mathcal{I}(a')}{\partial a_n'} na_{n-1}' = 0,$$

and hence the differential equation

$$0 = \frac{\partial \mathcal{I}(a')}{\partial a_0'} \frac{da_0'}{d\mu} + \frac{\partial \mathcal{I}(a')}{\partial a_1'} \frac{da_1'}{d\mu} + \cdots,$$

is also satisfied. Then, finally

$$\frac{d\mathcal{I}(a')}{d\mu} = 0,$$

thus $\mathcal{I}(a')$ is independent of μ, that is,

$$\mathcal{I}(a'_0, a'_1, \ldots, a'_n) = \mathcal{F}(a_0, a_1, \ldots, a_n),$$

for some function $\mathcal{F}$. For $\mu = 0$, one obtains from this

$$\mathcal{I}(a_0, a_1, \ldots, a_n) = \mathcal{F}(a_0, a_1, \ldots, a_n),$$

therefore

$$\mathcal{I}(a'_0, a'_1, \ldots, a'_n) = \mathcal{I}(a_0, a_1, \ldots, a_n)$$

is indeed an invariant with respect to (2). Thus, the differential equation (V) at the same time expresses everything that we can obtain through application of a transformation of type (2).

Since the differential equation (V) is obviously of great importance, it is advisable to introduce a special operational symbol for the operation performed by it; namely, we want to define:

$$\mathbf{D} = a_0 \frac{\partial}{\partial a_1} + 2a_1 \frac{\partial}{\partial a_2} + 3a_2 \frac{\partial}{\partial a_3} + \cdots + na_{n-1} \frac{\partial}{\partial a_n}.$$

Then we have proven the following:

Theorem *Each invariant $\mathcal{I}$ of a form satisfies the differential equation*

$$\mathbf{D}\mathcal{I} = 0. \qquad (V')$$

Lecture IX (May 11, 1897)

An entirely similar analysis can be made for covariants. The equation

$$\mathcal{C}(a'_0, a'_1, \ldots, a'_n; x'_1, x'_2) = \mathcal{C}(a_0, a_1, \ldots, a_n; x_1, x_2)$$

has to be satisfied identically if one sets

$$a'_i = a_i + \binom{i}{1} a_{i-1}\mu + \binom{i}{2} a_{i-2}\mu^2 + \cdots + a_0\mu^i$$

and, at the same time—in addition—

$$\begin{aligned} x'_1 &= x_1 - \mu x_2, \\ x'_2 &= x_2. \end{aligned}$$

We always regard the unprimed letters as the independent variables. Differentiation of the above equation with respect to μ gives

$$\frac{dC(a')}{d\mu} = 0,$$

that is,

$$\frac{\partial C(a')}{\partial a'_0} \cdot 0 + \frac{\partial C(a')}{\partial a'_1} a'_0 + \frac{\partial C(a')}{\partial a'_2} 2a'_1 + \cdots + na'_{n-1} \frac{\partial C(a')}{\partial a'_n} - \frac{\partial C(a')}{\partial x'_1} x'_2 = 0.$$

Since $C(a')$ depends on the a'_i, x'_i in the exact same way as C depends on the a_i, x_i, we have the following:

Theorem *Each covariant C of a form satisfies the differential equation*

$$\mathbf{D}C = x_2 \frac{\partial C}{\partial x_1}. \qquad (VI)$$

Furthermore, every polynomial function C in the a and x, which satisfies the differential equation (VI) $\mathbf{D}C = x_2 \frac{\partial C}{\partial x_1}$, is a covariant with respect to the transformation (2).

This theorem contains the previous one about invariants as a special case. Because we can again introduce primed letters instead of the unprimed ones, we obtain the differential equation $\frac{dC(a';x')}{d\mu} = 0$ by substituting the transformation formulas that follow from (2), that is,

$$C(a'_0, a'_1, \ldots, a'_n; x'_1, x'_2) = \mathcal{F}(a_0, a_1, \ldots, a_n; x_1, x_2).$$

Hence, for $\mu = 0$, we have:

$$C(a_0, a_1, \ldots, a_n; x_1, x_2) = \mathcal{F}(a_0, a_1, \ldots, a_n; x_1, x_2),$$

so that indeed

$$C(a'_0, a'_1, \ldots, a'_n; x'_1, x'_2) = C(a_0, a_1, \ldots, a_n; x_1, x_2)$$

is a covariant with respect to (2).

The examples we have already considered above serve to verify our theorem. For the invariant

$$\mathcal{I} = a_0 a_2 - a_1^2$$

of a quadratic form, we have indeed

$$\mathbf{D}\mathcal{I} = a_0(-2a_1) + 2a_1 a_0 = -2a_0 a_1 + 2a_0 a_1 = 0.$$

Furthermore, the base form f itself satisfies the differential equation

$$\mathbf{D}f = x_2 \frac{\partial f}{\partial x_1}.$$

This is because

$$\mathbf{D}f = a_0\binom{n}{1}x_1^{n-1}x_2 + \cdots + \binom{n}{k}ka_{k-1}x_1^{n-k}x_2^k + \cdots + na_{n-1}x_2^n,$$

and

$$x_2\frac{\partial f}{\partial x_1} = na_0x_1^{n-1}x_2 + \cdots + \binom{n}{k-1}a_{k-1}(n-k+1)x_1^{n-k}x_2^k$$
$$+ \cdots + na_{n-1}x_2^n.$$

From this and the identity

$$k\binom{n}{k} = (n-k+1)\binom{n}{k-1},$$

the assertion follows immediately.

The third type

$$\begin{aligned} x_1 &= x_1', \\ x_2 &= \nu x_1' + x_2' \end{aligned} \tag{3}$$

(with transformation determinant $\delta = 1$) admits similar observations as type (2). After all, only the role of the variables is interchanged. Type (3) transforms the form

$$f = a_0x_1^n + \binom{n}{1}a_1x_1^{n-1}x_2 + \cdots + \binom{n}{i}a_ix_1^{n-i}x_2^i + \cdots + a_nx_2^n$$

to

$$\begin{aligned} f &= a_0{x_1'}^n + \binom{n}{1}a_1{x_1'}^{n-1}(\nu x_1' + x_2') + \cdots \\ &\quad + \binom{n}{i}a_i{x_1'}^{n-i}(\nu x_1' + x_2')^i + \cdots + a_n(\nu x_1' + x_2')^n \\ &= a_0'{x_1'}^n + \binom{n}{1}a_1'{x_1'}^{n-1}x_2' + \cdots + \binom{n}{i}a_i'{x_1'}^{n-i}{x_2'}^i + \cdots + a_n'x_2', \end{aligned}$$

where

$$a_i' = a_i + \binom{n-i}{1}a_{i+1}\nu + \binom{n-i}{2}a_{i+2}\nu^2 + \cdots + a_n\nu^{n-i},$$

which is easily seen by applying the identity

$$\binom{n}{i+k}\binom{i+k}{i} = \binom{n}{i}\binom{n-i}{k}$$

to the general term. Furthermore, one has

$$\frac{da_i'}{d\nu} = (n-i)a_{i+1}'.$$

An invariant of f now has to make the equation

$$\mathcal{I}(a_0', a_1', \ldots, a_n') = \mathcal{I}(a_0, a_1, \ldots, a_n)$$

hold identically in the a, ν (and, later for covariants, the x). Differentiation with respect to ν produces:

$$\frac{d\mathcal{I}(a')}{d\nu} = 0,$$

or

$$0 = \frac{\partial\mathcal{I}(a')}{\partial a_0'}\frac{d\,a_0'}{d\,\nu} + \frac{\partial\mathcal{I}(a')}{\partial a_1'}\frac{d\,a_1'}{d\,\nu} + \cdots + \frac{\partial\mathcal{I}(a')}{\partial a_{n-1}'}\frac{d\,a_{n-1}'}{\partial\nu} + \frac{\partial\mathcal{I}(a')}{\partial a_n'}\frac{d\,a_n'}{d\,\nu},$$

or, according to the above formula,

$$na_1\frac{\partial\mathcal{I}}{\partial a_0} + (n-1)\,a_2\frac{\partial\mathcal{I}}{\partial a_1} + (n-2)\,a_3\frac{\partial\mathcal{I}}{\partial a_2} + \cdots + a_n\frac{\partial\mathcal{I}}{\partial a_{n-1}} = 0,$$

where the primed letters have been replaced by the unprimed ones. We introduce the symbol

$$\mathbf{\Delta} = na_1\frac{\partial}{\partial a_0} + (n-1)\,a_2\frac{\partial}{\partial a_1} + (n-2)\,a_3\frac{\partial}{\partial a_2} + \cdots + a_n\frac{\partial}{\partial a_{n-1}},$$

and obtain the following:

Theorem *Each invariant $\mathcal{I}$ of a form satisfies the differential equation*

$$\mathbf{\Delta}\mathcal{I} = 0. \qquad (VII)$$

Conversely, every polynomial function $\mathcal{I}$, which satisfies the differential equation $\mathbf{\Delta}\mathcal{I} = 0$, and which depends only on the a, is an invariant with respect to (3).

Finally, for the covariant

$$\mathcal{C}(a_0', a_1', \ldots, a_n';\, x_1', x_2') = \mathcal{C}(a_0, a_1, \ldots, a_n;\, x_1, x_2),$$

where

$$x_1' = x_1,$$
$$x_2' = \nu x_1 + x_2,$$

we must also have

$$\frac{d\,\mathcal{C}(a')}{d\,\nu} = 0,$$

that is,

$$\Delta C(a') - \frac{\partial C(a')}{\partial x_2'}\, x_1' = 0;$$

so we obtain the following:

Theorem *Each covariant of a form satisfies the differential equation*

$$\Delta C = x_1 \frac{\partial C}{\partial x_2}. \qquad (VIII)$$

And furthermore, we have:

Theorem *Every polynomial function of the a and the x, which satisfies the differential equation $\Delta C = x_1 \frac{\partial C}{\partial x_2}$, is a covariant with respect to the transformation (3).*

The examples again serve to verify these theorems: If

$$\mathcal{I} = a_0 a_2 - a_1^2,$$

then one has

$$\Delta \mathcal{I} = 2a_1 a_2 - 2a_2 a_1 = 0.$$

One can show equally easily that the form f itself satisfies the equation

$$\Delta f = x_1 \frac{\partial f}{\partial x_2}.$$

We have now already penetrated very deeply into invariant theory, even more so since we can now, at the conclusion of this section, formulate the important theorem whose proof now poses no more difficulties whatsoever:

Theorem *Every polynomial function of the coefficients $a_0, a_1, \ldots, a_n$ and the variables x_1, x_2 that is homogeneous in the a as well as the x and that possesses the properties derived in the present section is a covariant.*

The proof of this theorem is as follows. Let C be a co- or invariant that satisfies the equations (III), (IV), (VI), and (VIII). We transform the base form f via the general linear transformation:

$$x_1 = \alpha_{11}x_1' + \alpha_{12}x_2',$$
$$x_2 = \alpha_{21}x_1' + \alpha_{22}x_2'.$$

But we obtain the same result if, assuming $\alpha_{22} \neq 0$, we apply successively the transformations (1), (2), (3) with suitably chosen κ, λ, μ, ν. Thus we first introduce into C the $a'_{(1)}$, $x'_{(1)}$, as they are produced by type (1); then

$$C(a'_{(1)};\, x'_{(1)}) = \kappa^p \lambda^p C(a;\, x),$$

and therefore (Lecture V):

$$C(a'_{(1)};\, x'_{(1)}) = (\alpha_{11}\alpha_{22} - \alpha_{12}\alpha_{21})^p\, C(a;\, x).$$

Subsequently, we do the same with type (2), and then type (3). But with both of these no new factor appears, and so it indeed follows that:

$$C(a'_0, a'_1, \ldots, a'_n;\, x'_1, x'_2) = \delta^p C(a_0, a_1, \ldots, a_n;\, x_1, x_2).$$

But if $\alpha_{22} = 0$, then (Lecture V) one has to apply the transformation $(3')$, which introduces the factor $(-1)^p$. Nothing else can have changed, since $(3')$ is composed of (3), (2), (3), (1) for $\delta = -1$. But prior to this, one has to apply a transformation which, according to the above, introduces the factor $(\alpha_{12}\alpha_{21})^p$. Therefore, after applying all the transformations, one will obtain

$$\begin{aligned} C(a'_0, a'_1, \ldots, a'_n;\, x'_1, x'_2) &= (-1)^p (\alpha_{12}\alpha_{21})^p C(a_0, a_1, \ldots, a_n;\, x_1, x_2) \\ &= (-\alpha_{12}\alpha_{21})^p C(a_0, a_1, \ldots, a_n;\, x_1, x_2) \\ &= \delta^p C(a_0, a_1, \ldots, a_n;\, x_1, x_2). \end{aligned}$$

This completes the proof of the theorem.

With the last theorem, we have gained a new definition of in- and covariants. It enables us to recognize the invariant property of a given function much more easily than is possible through computation—as the simplest example of the invariant of a quadratic base form already shows.

Lecture X (May 13, 1897)

I.5 The operation symbols D and Δ

The necessary and sufficient conditions for the invariants and covariants, which we derived in the preceding section, can be reduced to a smaller number. This is done using certain relations of the symbols **D** and **Δ**, which we have to derive in this section. The symbols **D**, **Δ** obey the simplest laws of ordinary differentiation. As is immediately obvious, one has the formulas:

$$\mathbf{D}(\mathcal{A}_1+\mathcal{A}_2)=\mathbf{D}(\mathcal{A}_1)+\mathbf{D}(\mathcal{A}_2); \qquad \boldsymbol{\Delta}(\mathcal{A}_1+\mathcal{A}_2)=\boldsymbol{\Delta}\mathcal{A}_1+\boldsymbol{\Delta}\mathcal{A}_2;$$
$$\mathbf{D}(\mathcal{A}_1\mathcal{A}_2)=\mathcal{A}_1\mathbf{D}\mathcal{A}_2+\mathcal{A}_2\mathbf{D}\mathcal{A}_1; \qquad \boldsymbol{\Delta}(\mathcal{A}_1\mathcal{A}_2)=\mathcal{A}_1\boldsymbol{\Delta}\mathcal{A}_2+\mathcal{A}_2\boldsymbol{\Delta}\mathcal{A}_1;$$
$$\mathbf{D}(\text{const})=0; \qquad \boldsymbol{\Delta}(\text{const})=0;$$
$$\mathbf{D}(\text{const}\cdot\mathcal{A})=\text{const}\cdot\mathbf{D}\mathcal{A}; \qquad \boldsymbol{\Delta}(\text{const}\cdot\mathcal{A})=\text{const}\cdot\boldsymbol{\Delta}\mathcal{A},$$

where $\mathcal{A}$, $\mathcal{A}_1$, and $\mathcal{A}_2$ denote arbitrary functions.

We furthermore want to observe the effect of the operations $\mathbf{D}$ and $\boldsymbol{\Delta}$ on the degree and the weight of a function. Let

$$\mathcal{A}=\sum Z_{\nu_0\ldots\nu_n}a_0^{\nu_0}a_1^{\nu_1}\cdots a_n^{\nu_n}$$

be a homogeneous, isobaric function of the a (the Z should not contain the a, but need not be numbers). According to the preceding we only need to consider

$$\mathbf{D}(a_0^{\nu_0}a_1^{\nu_1}\cdots a_n^{\nu_n}) \qquad \text{and} \qquad \boldsymbol{\Delta}(a_0^{\nu_0}\cdots a_n^{\nu_n}).$$

We obviously have

$$\begin{aligned}\mathbf{D}(a_0^{\nu_0}a_1^{\nu_1}\cdots a_n^{\nu_n})&=\sum_{i=1}^{n} ia_{i-1}\frac{\partial(a_0^{\nu_0}a_1^{\nu_1}\cdots a_n^{\nu_n})}{\partial a_i}\\&=\sum_{i=1}^{n} ia_{i-1}\nu_i a_0^{\nu_0}a_1^{\nu_1}\cdots a_{i-1}^{\nu_{i-1}}a_{i+1}^{\nu_{i+1}}\cdots a_n^{\nu_n}a_i^{\nu_i-1},\end{aligned}$$

that is, the degree of a term is

$$1+\nu_0+\nu_1+\cdots+\nu_{i-1}+\nu_{i+1}+\cdots+\nu_n+\nu_i-1=g,$$

and the weight is

$$\begin{aligned}&\nu_1+2\nu_2+\cdots+(i-1)(\nu_{i-1}+1)+i(\nu_i-1)+(i+1)\nu_{i+1}+\cdots+n\nu_n\\&\quad=\nu_1+2\nu_2+\cdots+(i-1)\nu_{i-1}+i\nu_i+(i+1)\nu_{i+1}+\cdots\\&\qquad+n\nu_n+i-1-i\\&\quad=p-1.\end{aligned}$$

Likewise

$$\begin{aligned}\boldsymbol{\Delta}(a_0^{\nu_0}a_1^{\nu_1}\cdots a_n^{\nu_n})&=\sum_{i=0}^{n-1}(n-i)a_{i+1}\frac{\partial(a_0^{\nu_0}a_1^{\nu_1}\cdots a_n^{\nu_n})}{\partial a_i}\\&=\sum_{i=0}^{n-1}(n-i)a_{i+1}\nu_i a_i^{\nu_i-1}a_0^{\nu_0}\cdots a_{i-1}^{\nu_{i-1}}a_{i+1}^{\nu_{i+1}}\cdots a_n^{\nu_n}.\end{aligned}$$

Therefore, the degree of a term is

$$1 + \nu_0 + \nu_1 + \cdots + \nu_{i-1} + \nu_{i+1} + \cdots + \nu_n + \nu_i - 1 = g,$$

and the weight is

$$\begin{aligned}
&\nu_1 + 2\nu_2 + \cdots + (i-1)\nu_{i-1} + i(\nu_i - 1) + (i+1)(\nu_{i+1} + 1) + \cdots + n\nu_n \\
&\quad = \nu_1 + 2\nu_2 + \cdots + (i-1)\nu_{i-1} + i\nu_i + (i+1)\nu_{i+1} + \cdots \\
&\qquad + n\nu_n - i + i + 1 \\
&\quad = p + 1.
\end{aligned}$$

We therefore have the following:

Theorem *The operations* $\mathbf{D}$ *and* $\boldsymbol{\Delta}$, *applied to a homogeneous, isobaric function, leave the degree* g *in the* a_i *unchanged;* $\mathbf{D}$ *lowers the weight* p *by 1, that is, to* $p-1$*;* $\boldsymbol{\Delta}$ *raises it by 1, to* $p+1$. *Homogeneity and isobarity are, therefore, preserved by these operations.*

After these preparations we prove the following:

Theorem *If* $\mathcal{A}$ *is a homogeneous and isobaric function in the* a_i, *of degree* g *and weight* p, *then*

$$(\mathbf{D}\boldsymbol{\Delta} - \boldsymbol{\Delta}\mathbf{D})\mathcal{A} = (ng - 2p)\mathcal{A}. \qquad (I)$$

We prove this theorem in three steps:

1. If formula (I) is valid for two expressions $\mathcal{A}_1$ and $\mathcal{A}_2$ which have the same degree g and the same weight p, then it is also valid for the sum $\mathcal{A}_1 + \mathcal{A}_2$. From

$$\begin{aligned}
(\mathbf{D}\boldsymbol{\Delta} - \boldsymbol{\Delta}\mathbf{D})\mathcal{A}_1 &= (ng - 2p)\mathcal{A}_1, \\
(\mathbf{D}\boldsymbol{\Delta} - \boldsymbol{\Delta}\mathbf{D})\mathcal{A}_2 &= (ng - 2p)\mathcal{A}_2
\end{aligned}$$

it follows immediately by addition that

$$(\mathbf{D}\boldsymbol{\Delta} - \boldsymbol{\Delta}\mathbf{D})(\mathcal{A}_1 + \mathcal{A}_2) = (ng - 2p)(\mathcal{A}_1 + \mathcal{A}_2).$$

2. If the formula is valid for $\mathcal{A}_1$ (degree $= g_1$, weight $= p_1$) and $\mathcal{A}_2$ (degree $= g_2$, weight $= p_2$), then it is also valid for the product $\mathcal{A}_1\mathcal{A}_2$. The product is easily seen to be a homogeneous and isobaric function of degree $g = g_1 + g_2$ and weight $p = p_1 + p_2$. According to our assumptions, we have now

$$\begin{aligned}
(\mathbf{D}\boldsymbol{\Delta} - \boldsymbol{\Delta}\mathbf{D})\mathcal{A}_1 &= (ng_1 - 2p_1)\mathcal{A}_1, \\
(\mathbf{D}\boldsymbol{\Delta} - \boldsymbol{\Delta}\mathbf{D})\mathcal{A}_2 &= (ng_2 - 2p_2)\mathcal{A}_2.
\end{aligned}$$

Using the rules stated above one finds for the product:

$$
\begin{aligned}
(\mathbf{D\Delta} - \mathbf{\Delta D})(\mathcal{A}_1\mathcal{A}_2) \\
&= \mathbf{D\Delta}(\mathcal{A}_1\mathcal{A}_2) - \mathbf{\Delta D}(\mathcal{A}_1\mathcal{A}_2) \\
&= \mathbf{D}\{\mathcal{A}_1\mathbf{\Delta}\mathcal{A}_2 + \mathcal{A}_2\mathbf{\Delta}\mathcal{A}_1\} - \mathbf{\Delta}\{\mathcal{A}_1\mathbf{D}\mathcal{A}_2 + \mathcal{A}_2\mathbf{D}\mathcal{A}_1\} \\
&= \mathcal{A}_1\mathbf{D\Delta}\mathcal{A}_2 + \{\mathbf{\Delta}\mathcal{A}_2\}\{\mathbf{D}\mathcal{A}_1\} + \mathcal{A}_2\mathbf{D\Delta}\mathcal{A}_1 + \{\mathbf{\Delta}\mathcal{A}_1\}\{\mathbf{D}\mathcal{A}_2\} \\
&\quad - \mathcal{A}_1\mathbf{\Delta D}\mathcal{A}_2 - \{\mathbf{D}\mathcal{A}_2\}\{\mathbf{\Delta}\mathcal{A}_1\} - \mathcal{A}_2\mathbf{\Delta D}\mathcal{A}_1 - \{\mathbf{D}\mathcal{A}_1\}\{\mathbf{\Delta}\mathcal{A}_2\} \\
&= \mathcal{A}_1\{\mathbf{D\Delta}\mathcal{A}_2 - \mathbf{\Delta D}\mathcal{A}_2\} + \mathcal{A}_2\{\mathbf{D\Delta}\mathcal{A}_1 - \mathbf{\Delta D}\mathcal{A}_1\} \\
&= (ng_2 - 2p_2)\mathcal{A}_1\mathcal{A}_2 + (ng_1 - 2p_1)\mathcal{A}_2\mathcal{A}_1 \\
&= \big(n(g_1 + g_2) - 2(p_1 + p_2)\big)\mathcal{A}_1\mathcal{A}_2;
\end{aligned}
$$

therefore, indeed

$$(\mathbf{D\Delta} - \mathbf{\Delta D})(\mathcal{A}_1\mathcal{A}_2) = (ng - 2p)\mathcal{A}_1\mathcal{A}_2.$$

3. The formula is valid for $\mathcal{A} =$ const., since then it only says $0 = 0$, and for

$$\mathcal{A} = a_i,$$

because in this case

$$
\begin{aligned}
\mathbf{D\Delta}a_i &= \mathbf{D}(n - i)a_{i+1} = (n - i)(i + 1)a_i, \\
\mathbf{\Delta D}a_i &= \mathbf{\Delta}ia_{i-1} = i(n - i + 1)a_i.
\end{aligned}
$$

Consequently,

$$
\begin{aligned}
(\mathbf{D\Delta} - \mathbf{\Delta D})a_i &= (ni - i^2 + n - i - in + i^2 - i)\, a_i \\
&= (n - 2i)\, a_i\,.
\end{aligned}
$$

Finally the general theorem follows directly from these three facts.
As an example, consider for $n = 3$

$$\mathcal{A} = a_1a_3 + 5a_2^2; \qquad g = 2,\ p = 4.$$

Here we have

$$
\begin{aligned}
\mathbf{D} &= a_0\frac{\partial}{\partial a_1} + 2a_1\frac{\partial}{\partial a_2} + 3a_2\frac{\partial}{\partial a_3}, \\
\mathbf{\Delta} &= 3a_1\frac{\partial}{\partial a_0} + 2a_2\frac{\partial}{\partial a_1} + a_3\frac{\partial}{\partial a_2},
\end{aligned}
$$

hence

$$\begin{aligned}
\mathbf{D}\mathcal{A} &= a_0a_3 + 2a_1 \cdot 10a_2 + 3a_2a_1 = a_0a_3 + 23a_1a_2,\\
\mathbf{\Delta D}\mathcal{A} &= 3a_1a_3 + 2a_2 \cdot 23a_2 + a_3 \cdot 23a_1 = 26a_1a_3 + 46a_2^2,\\
\mathbf{\Delta}\mathcal{A} &= 3a_1 \cdot 0 + 2a_2a_3 + a_3 \cdot 10a_2 = 12a_2a_3,\\
\mathbf{D\Delta}\mathcal{A} &= a_0 \cdot 0 + 2a_1 \cdot 12a_3 + 3a_2 \cdot 12a_2 = 24a_1a_3 + 36a_2^2.
\end{aligned}$$

From this it follows that

$$\begin{aligned}
(\mathbf{D\Delta} - \mathbf{\Delta D})\mathcal{A} &= -2a_1a_3 - 10a_2^2 = -2(a_1a_3 + 5a_2^2)\\
&= -2\mathcal{A},
\end{aligned}$$

as asserted.

We shall need to derive two general formulas from formula (I). Let $\mathcal{A}$ be a homogeneous, isobaric function of degree g and weight p. Then $\mathbf{D}\mathcal{A}$ is a homogeneous and isobaric function of degree g and weight $p-1$, and generally, if we let

$$\mathbf{D}^2 = \mathbf{DD}, \quad \mathbf{D}^3 = \mathbf{DDD}, \ldots,$$

then $\mathbf{D}^k\mathcal{A}$ is a homogeneous isobaric function of degree g and weight $p-k$. If we now apply the operation $\mathbf{D}$ to both sides of formula (I), then we obtain

$$(\mathbf{D}^2\mathbf{\Delta} - \mathbf{D\Delta D})\mathcal{A} = (ng - 2p)\mathbf{D}\mathcal{A}.$$

Application of formula (I) to $\mathbf{D}\mathcal{A}$, on the other hand, results in

$$(\mathbf{D\Delta D} - \mathbf{\Delta D}^2)\mathcal{A} = (ng - 2p + 2)\mathbf{D}\mathcal{A}.$$

Addition of both formulas gives (omitting the variable):

$$\mathbf{D}^2\mathbf{\Delta} - \mathbf{\Delta D}^2 = (2ng - 4p + 2)\mathbf{D},$$

or

$$\mathbf{D}^2\mathbf{\Delta} - \mathbf{\Delta D}^2 = 2(ng - 2p + 1)\mathbf{D}.$$

But if one applies the operation $\mathbf{\Delta}$ to (I), then it follows that

$$(\mathbf{\Delta D\Delta} - \mathbf{\Delta}^2\mathbf{D})\mathcal{A} = (ng - 2p)\mathbf{\Delta}\mathcal{A};$$

and if one applies formula (I) to $\mathbf{\Delta}\mathcal{A}$, then one obtains

$$(\mathbf{D\Delta}^2 - \mathbf{\Delta D\Delta})\mathcal{A} = (ng - 2p - 2)\mathbf{\Delta}\mathcal{A},$$

and addition of these formulas results in

$$(\mathbf{D\Delta}^2 - \mathbf{\Delta}^2\mathbf{D}) = 2(ng - 2p - 1)\mathbf{\Delta}.$$

Lecture XI (May 14, 1897)

The following two important formulas are valid in general:

$$\mathbf{D}^k\boldsymbol{\Delta} - \boldsymbol{\Delta}\mathbf{D}^k = k(ng - 2p + k - 1)\mathbf{D}^{k-1}, \qquad (II)$$

$$\mathbf{D}\boldsymbol{\Delta}^k - \boldsymbol{\Delta}^k\mathbf{D} = k(ng - 2p - k + 1)\boldsymbol{\Delta}^{k-1}. \qquad (III)$$

This is easily proven by induction from k to $k+1$. Namely, suppose the formulas (II), (III) are valid for k. We first apply the operation $\mathbf{D}$ to (II), then formula (I) to $\mathbf{D}^k$; likewise, we first apply the operation $\boldsymbol{\Delta}$ to (III), then formula (I) to $\boldsymbol{\Delta}^k$. Then we obtain the following identities:

$$\begin{aligned}
&\mathbf{D}^{k+1}\boldsymbol{\Delta} - \mathbf{D}\boldsymbol{\Delta}\mathbf{D}^k = k(ng - 2p + k - 1)\mathbf{D}^k,\\
&\mathbf{D}\boldsymbol{\Delta}\mathbf{D}^k - \boldsymbol{\Delta}\mathbf{D}^{k+1} = \big(ng - 2(p-k)\big)\mathbf{D}^k,\\
&\boldsymbol{\Delta}\mathbf{D}\boldsymbol{\Delta}^k - \boldsymbol{\Delta}^{k+1}\mathbf{D} = k(ng - 2p - k + 1)\boldsymbol{\Delta}^k,\\
&\mathbf{D}\boldsymbol{\Delta}^{k+1} - \boldsymbol{\Delta}\mathbf{D}\boldsymbol{\Delta}^k = \big(ng - 2(p+k)\big)\boldsymbol{\Delta}^k.
\end{aligned}$$

And if one adds the first two formulas on the one hand and the second two on the other hand, then one obtains the formulas (II) and (III) for $k+1$. But since they are valid for $k = 2$, they are valid in general. For $k = 1$, they both turn into formula (I), provided one defines

$$\mathbf{D}^0\mathcal{A} = \boldsymbol{\Delta}^0\mathcal{A} = \mathcal{A}.$$

We shall assume that the formulas we derived are always being applied to a homogeneous, isobaric expression.

I.6 The smallest system of conditions for the determination of the invariants and covariants

Using the relations derived in the previous section, we can now substantially simplify the necessary and sufficient conditions of Section I.4 for in- and covariants. We first show this for the invariants.

Let $\mathcal{A}$ be a homogeneous isobaric function of the a_i of degree g and weight p, where $ng = 2p$. Suppose further that $\mathcal{A}$ satisfies the differential equation $\mathbf{D}\mathcal{A} = 0$. Then we want to show that $\mathcal{A}$ also satisfies the differential equation $\boldsymbol{\Delta}\mathcal{A} = 0$, and hence is an invariant.

We form the sequence of expressions:

	$\mathcal{A}$	$\mathbf{\Delta}\mathcal{A}$	$\mathbf{\Delta}^2\mathcal{A}$	$\mathbf{\Delta}^3\mathcal{A}$	$\dots$	$\mathbf{\Delta}^{k-1}\mathcal{A}$	$\mathbf{\Delta}^k\mathcal{A}$	$\dots$
degree	g	g	g	g	$\dots$	g	g	$\dots$
weight	p	$p+1$	$p+2$	$p+3$	$\dots$	$p+k-1$	$p+k$	$\dots$

But, since $\mathbf{\Delta}^k\mathcal{A}$ has degree g for all k, the weight cannot take on arbitrary values; therefore, from some $\mathbf{\Delta}^i$ on, all have to be identically zero. Let this be $\mathbf{\Delta}^k\mathcal{A} = 0$, while $\mathbf{\Delta}^{k-1}\mathcal{A} \neq 0$. But, according to formula (III) in the preceding section, we have

$$\mathbf{D}\mathbf{\Delta}^k\mathcal{A} - \mathbf{\Delta}^k\mathbf{D}\mathcal{A} = k(-k+1)\mathbf{\Delta}^{k-1}\mathcal{A}.$$

Therefore, we must have

$$k(k-1)\mathbf{\Delta}^{k-1}\mathcal{A} = 0.$$

Here it can neither happen that $\mathbf{\Delta}^{k-1}\mathcal{A} = 0$ (because of our assumption), nor that $k = 0$, because then $\mathcal{A}$ would be identically zero, and so it follows that

$$k = 1,$$

that is,

$$\mathbf{\Delta}\mathcal{A} = 0,$$

and our assertion is proven. Herewith, we have the following:

Theorem *Every homogeneous isobaric function $\mathcal{I}$ of the coefficients $a_0, a_1, \dots, a_n$, of degree g and weight p, where $ng = 2p$, is an invariant if it satisfies the differential equation $\mathbf{D}\mathcal{I} = 0$.*

This theorem already enables us to construct as many invariants as we wish. We give some examples.

1. $n = 1$; $\quad f = a_0x_1 + a_1x_2$.

Here an invariant must have the form:

$$\mathcal{I} = \sum Z_{\nu_0\nu_1} a_0^{\nu_0} a_1^{\nu_1},$$

so it follows that

$$\nu_0 + \nu_1 = g, \qquad \nu_1 = p,$$

$$2\nu_1 = \nu_0 + \nu_1,$$

that is,

$$\nu_1 = \nu_0.$$

Thus, up to a constant, one has

$$\mathcal{I} = (a_0 a_1)^\nu,$$

or, more simply,

$$\mathcal{I} = a_0 a_1.$$

But this is not an invariant, since the differential equation

$$a_0^2 = \mathbf{D}\mathcal{I} = 0$$

is not satisfied. Therefore, "a linear form has no invariants," which can be seen directly as well.

2. $n = 2$; $\qquad f = a_0 x_1^2 + 2a_1 x_1 x_2 + a_2 x_2^2.$

(a) From $g = 1$ it follows that $p = 1$; so one would have

$$\mathcal{I} = a_1,$$

and this is not an invariant, since the differential equation is not satisfied. Consequently, a second order form does not have an invariant of degree one.

(b) From $g = 2$ follows $p = 2$. So one must have

$$\mathcal{I} = Z a_0 a_2 + Z' a_1^2.$$

In order for the differential equation $\mathbf{D}\mathcal{I} = 0$ to be satisfied

$$\mathbf{D}\mathcal{I} = a_0 Z' 2a_1 + 2a_1 Z a_0 = 0,$$

or

$$2a_0 a_1 (Z + Z') = 0,$$

thus

$$Z' = -Z$$

must hold. Therefore, one has the following:

Theorem *A quadratic form has only one invariant of degree two, namely,*

$$a_0 a_2 - a_1^2.$$

We always omit constant factors, which can of course be added at any time.

3. $n = 4$, $g = 2$ results in $p = 4$. We then set

$$\mathcal{I}_2 = Z a_0 a_4 + Z' a_1 a_3 + Z'' a_2^2.$$

Since we have

$$\begin{aligned} \mathbf{D}a_0a_4 &= 4a_0a_3, \\ \mathbf{D}a_1a_3 &= a_0a_3 + 3a_1a_2, \\ \mathbf{D}a_2^2 &= 4a_1a_2, \end{aligned}$$

it follows that

$$\mathbf{D}\mathcal{I} = 4Za_0a_3 + Z'a_0a_3 + 3Z'a_1a_2 + Z''4a_1a_2 = 0$$

or

$$(4Z + Z')a_0a_3 + (3Z' + 4Z'')a_1a_2 = 0.$$

This equation has to be identically satisfied, so it follows that

$$Z' = -4Z, \qquad 4Z'' = -3Z' = 12Z,$$

and hence

$$Z'' = 3Z.$$

Hence there is only one quadratic invariant of the biquadratic form, which is

$$\mathcal{I}_2 = a_0a_4 - 4a_1a_3 + 3a_2^2.$$

The equation $\mathbf{\Delta}\mathcal{I} = 0$ is satisfied since

$$\mathbf{\Delta}\mathcal{I}_2 = 4a_1a_4 - 3a_2 \cdot 4a_3 + 2a_3 \cdot 6a_2 - a_4 \cdot 4a_1 = 0.$$

Lecture XII (May 17, 1897)

4. $n = 4, \quad g = 3$ results in $p = 6$. Thus, we set

$$\mathcal{I}_3 = Za_0a_2a_4 + Z'a_0a_3^2 + Z''a_1^2a_4 + Z'''a_1a_2a_3 + Z^{IV}a_2^3,$$

and it only remains to satisfy the differential equation $\mathbf{D}\mathcal{I} = 0$. We obtain:

$$\begin{aligned} \mathbf{D}\mathcal{I}_3 = {} & a_0\{2Z''a_1a_4 + Z'''a_2a_3\} + 2a_1\{Za_0a_4 + Z'''a_1a_3 + 3Z^{IV}a_2^2\} \\ & + 3a_2\{2Z'a_0a_3 + Z'''a_1a_2\} + 4a_3\{Za_0a_2 + Z''a_1^2\} \\ = {} & 0. \end{aligned}$$

Thus we must have

$$\begin{aligned} 2Z'' + 2Z &= 0, & 2Z''' + 4Z'' &= 0, \\ Z''' + 6Z' + 4Z &= 0, & 6Z^{IV} + 3Z''' &= 0, \end{aligned}$$

from which we get

$$Z' = -Z, \qquad Z'' = -Z, \qquad Z''' = 2Z, \qquad Z^{IV} = -Z.$$

Hence, there is only one invariant of degree three of a biquadratic form, which is

$$\mathcal{I}_3 = a_0a_2a_4 - a_0a_3^2 - a_1^2a_4 + 2a_1a_2a_3 - a_2^3.$$

The differential equation $\mathbf{\Delta}\mathcal{I}_3 = 0$ is automatically satisfied.

Likewise, we must now reduce the number of conditions for a covariant of a form to be as small as possible. As before, let

$$\mathcal{C} = C_0x_1^m + \binom{m}{1}C_1x_1^{m-1}x_2 + \binom{m}{2}C_2x_1^{m-2}x_2^2 + \cdots + C_mx_2^m;$$

this expression $\mathcal{C}$ is a covariant of the form $f^{(n)}$ if and only if the following conditions are satisfied: $C_0, C_1, \ldots, C_m$ are homogeneous functions of equal degree g, as well as isobaric expressions of weight p, $p+1$, $p+2$, $\ldots, p+m$, respectively, where $ng - m = 2p$ must hold, and, furthermore, the differential equations

$$\mathbf{D}\mathcal{C} = x_2\,\frac{\partial\mathcal{C}}{\partial x_1}, \qquad \mathbf{\Delta}\mathcal{C} = x_1\,\frac{\partial\mathcal{C}}{\partial x_2}$$

are identically satisfied.

Each of these two differential equations can be replaced by $m+1$ others. The first one is obviously equivalent to the following:

$$\mathbf{D}C_0x_1^m + \binom{m}{1}\mathbf{D}C_1x_1^{m-1}x_2 + \cdots + \binom{m}{i}\mathbf{D}C_ix_1^{m-i}x_2^i + \cdots + \mathbf{D}C_mx_2^m$$
$$= mC_0x_1^{m-1}x_2 + \cdots + \binom{m}{i-1}C_{i-1}(m-i+1)x_1^{m-i}x_2^i + \cdots$$
$$+ \binom{m}{1}C_{m-1}x_2^m.$$

Since this equation has to be identically satisfied, each of the following $m+1$ differential equations has to hold:

$$\begin{aligned} \mathbf{D}C_0 &= 0, \\ \mathbf{D}C_1 &= C_0, \\ \mathbf{D}C_2 &= 2C_1, \\ &\cdots \\ \mathbf{D}C_i &= iC_{i-1}, \\ &\cdots \\ \mathbf{D}C_m &= mC_{m-1}. \end{aligned}$$

We treat the second differential equation in the same way. Then we obtain

$$\begin{aligned}\Delta C_0 x_1^m + \binom{m}{1}\Delta C_1 x_1^{m-1}x_2 + \cdots + \binom{m}{i}\Delta C_i x_1^{m-i}x_2^i + \cdots + \Delta C_m x_2^m \\ = \binom{m}{1}C_1 x_1^m + 2\binom{m}{2}C_2 x_1^{m-1}x_2 + \cdots + (i+1)\binom{m}{i+1}C_{i+1}x_1^{m-i}x_2^i \\ + \cdots + mC_m x_2^{m-1}x_1,\end{aligned}$$

and, consequently, the following differential equations must hold individually, from which we can deduce in turn the validity of the differential equation $\Delta C = x_1 \frac{\partial C}{\partial x_2}$:

$$\begin{aligned}\Delta C_0 &= mC_1,\\ \Delta C_1 &= (m-1)C_2,\\ &\cdots\\ \Delta C_i &= (m-i)C_{i+1},\\ &\cdots\\ \Delta C_{m-1} &= C_m,\\ \Delta C_m &= 0.\end{aligned}$$

The first m of these differential equations are of particular importance; they imply the following m equivalent ones, which in turn can replace them:

$$\begin{aligned}C_1 &= \frac{1}{m}\Delta C_0,\\ C_2 &= \frac{1}{m(m-1)}\Delta^2 C_0,\\ C_3 &= \frac{1}{m(m-1)(m-2)}\Delta^3 C_0,\\ &\cdots\\ C_i &= \frac{1}{m(m-1)\cdots(m-i+1)}\Delta^i C_0,\\ &\cdots\\ C_m &= \frac{1}{m(m-1)\cdots 2\cdot 1}\Delta^m C_0.\end{aligned}$$

We have herewith shown that the covariant is completely determined by its first term. For this reason, Sylvester has called the first coefficient

C_0 the *source of the covariant.* This source has to satisfy the differential equation

$$\mathbf{D}C_0 = 0.$$

But, as we will show promptly, with these $m+1$ differential equations, all the others are automatically satisfied, a fact that we want to record as the following theorem.

Lecture XIII (May 18, 1897)

Theorem *The expression*

$$C = C_0 x_1^m + \binom{m}{1} C_1 x_1^{m-1} x_2 + \binom{m}{2} C_2 x_1^{m-2} x_2^2 + \cdots + C_m x_2^m$$

is a covariant of the form $f^{(n)}(x_1, x_2)$ if and only if C_0 is a homogeneous isobaric function in the a, of degree g and weight p, such that $m = ng - 2p$, which satisfies the differential equation

$$\mathbf{D}C_0 = 0,$$

and if, furthermore, the $C_1, C_2, \ldots, C_m$ are derived from C_0 via the formulas

$$C_1 = \frac{1}{m} \mathbf{\Delta} C_0,$$

$$C_2 = \frac{1}{m(m-1)} \mathbf{\Delta}^2 C_0,$$

$$\cdots$$

$$C_i = \frac{1}{m(m-1)\cdots(m-i+1)} \mathbf{\Delta}^i C_0,$$

$$\cdots$$

$$C_m = \frac{1}{m(m-1)\cdots 2} \mathbf{\Delta}^m C_0.$$

These conditions are not only necessary but also sufficient. The hypotheses of homogeneity and isobarity are satisfied for all C (Lecture X). It therefore only remains to show that the differential equations $\mathbf{D}C_1 = C_0, \ldots, \mathbf{D}C_m = m\,C_{m-1}$, $\mathbf{\Delta}C_m = 0$ are also satisfied. Indeed, according to Section I.5, equation (III), we have

$$\mathbf{D}\mathbf{\Delta}^k C_0 - \mathbf{\Delta}^k \mathbf{D} C_0 = k(ng - 2p - k + 1)\mathbf{\Delta}^{k-1} C_0;$$

hence, according to our assumptions,

$$\mathbf{D}C_k m(m-1)\cdots(m-k+1) = k(m-k+1)C_{k-1}m(m-1)\cdots(m-k+2),$$

that is,

$$\mathbf{D}C_k = k\,C_{k-1}.$$

Finally, it remains to prove that $\mathbf{\Delta}C_m = 0$, or that

$$\mathbf{\Delta}^{m+1}C_0 = 0.$$

The sequence

$$C_0,\ \mathbf{\Delta}C_0, \ldots, \mathbf{\Delta}^{k-1}C_0,\ \mathbf{\Delta}^k C_0, \ldots$$

has to stop somewhere, because the weight cannot become arbitrarily large; say $\mathbf{\Delta}^{k-1}C_0 \neq 0$, $\mathbf{\Delta}^k C_0 = 0$. Then (Section I.5, equation (III))

$$\mathbf{D}\mathbf{\Delta}^k C_0 - \mathbf{\Delta}^k \mathbf{D} C_0 = k(m-k+1)\mathbf{\Delta}^{k-1}C_0,$$

and therefore

$$k(m-k+1) = 0.$$

Then, since $k = 0$ is not possible, it follows that

$$k = m+1.$$

Thus, we have indeed

$$\mathbf{\Delta}^{m+1}C_0 = 0.$$

This completes the proof of our theorem.

This theorem gives us again the ability to construct as many covariants as we wish. We demonstrate this through some examples.

1. The base form itself. Let

$$C = a_0 x_1^n + \binom{n}{1} a_1 x_1^{n-1} x_2 + \cdots + a_n x_2^n.$$

Indeed, $C_0 = a_0$ is of degree $g = 1$, homogeneous, and isobaric of weight $p = 0$, so that $n = n \cdot 1 - 2 \cdot 0$, and satisfies the differential equation

$$\mathbf{D}C_0 = \mathbf{D}a_0 = 0.$$

Furthermore,

$$a_1 = \frac{1}{n}\mathbf{\Delta} a_0 = \frac{1}{n} n a_1,$$

$$a_2 = \frac{1}{n(n-1)}\mathbf{\Delta}^2 a_0 = \frac{1}{n(n-1)}(n-1)n a_2,$$

$$\cdots$$

$$a_n = \frac{1}{n!}\mathbf{\Delta}^n a_0 = \frac{1}{n!} n!\, a_n.$$

2. $n = 3$. From $g = 2$, $p = 2$, it follows that $m = 2$. We set

$$C_0 = Z a_0 a_2 + Z' a_1^2 ;$$

then from the above we have

$$C_0 = a_0 a_2 - a_1^2,$$

from which we find that

$$C_1 = \frac{1}{2}\mathbf{\Delta} C_0 = \frac{1}{2}\big(3a_1 a_2 + 2a_2(-2a_1) + a_3 a_0\big)$$

$$= \frac{1}{2}(a_0 a_3 - a_1 a_2),$$

$$C_2 = \frac{1}{2}\mathbf{\Delta}^2 C_0 = \frac{1}{2}(3a_1 a_3 - 2a_2^2 - a_1 a_3)$$

$$= a_1 a_3 - a_2^2.$$

Hence, there is only one covariant of a cubic form of degree 2 and weight 2, which is

$$\mathcal{C} = (a_0 a_2 - a_1^2) x_1^2 + (a_0 a_3 - a_1 a_2) x_1 x_2 + (a_1 a_3 - a_2^2) x_2^2.$$

3. $n = 3$; $g = 3$, $p = 3$, and hence $m = 3$. We set

$$C_0 = Z a_0^2 a_3 + Z' a_0 a_1 a_2 + Z'' a_1^3,$$

and

$$\mathbf{D} C_0 = a_0 (Z' a_0 a_2 + 3Z'' a_1^2) + 2a_1 Z' a_0 a_1 + 3a_2 Z a_0^2 = 0$$

must hold, that is,

$$Z' + 3Z = 0, \qquad Z' = -3Z,$$
$$3Z'' + 2Z' = 0, \qquad Z'' = -\tfrac{2}{3}Z' = 2Z.$$

Therefore, we have

$$C_0 = a_0^2 a_3 - 3a_0 a_1 a_2 + 2a_1^3,$$

from which it follows that

$$\begin{aligned}\Delta C_0 &= 3(a_0 a_1 a_3 + a_1^2 a_2 - 2a_0 a_2^2),\\ \Delta^2 C_0 &= 6(2a_1^2 a_3 - a_1 a_2^2 - a_0 a_2 a_3),\\ \Delta^3 C_0 &= 6(3a_1 a_2 a_3 - 2a_2^3 - a_0 a_3^2).\end{aligned}$$

Consequently, there is only one covariant of a cubic form of degree 3 and weight 3, which is

$$\begin{aligned}\mathcal{C} = {}& (a_0^2 a_3 - 3a_0 a_1 a_2 + 2a_1^3)x_1^3 + 3(a_0 a_1 a_3 + a_1^2 a_2 - 2a_0 a_2^2)x_1^2 x_2\\ &- 3(a_0 a_2 a_3 - 2a_1^2 a_3 + a_1 a_2^2)x_1 x_2^2 - (a_0 a_3^2 - 3a_1 a_2 a_3 + 2a_2^3)x_2^3.\end{aligned}$$

It is also easily seen that the condition $\Delta^{m+1} C_0 = 0$, a consequence of our initial assumptions, is also satisfied.

We add some remarks to the first sections. We could have formulated the definition of invariants with somewhat greater generality from the beginning, without changing or generalizing the concept.

Definition *An invariant of a form f is a polynomial function of the coefficients $a_0, a_1, \dots, a_n$ that, under substitution of the transformed coefficients a_i' for the a_i, is transformed into a function of the a_i' which contains the original one as a factor.*

That is, we require that

$$\mathcal{I}(a_0', a_1', \dots, a_n') = \phi(\alpha) \cdot \mathcal{I}(a_0, a_1, \dots, a_n),$$

where $\phi(\alpha)$ is a polynomial function of the matrix entries α which does not depend on the a anymore, since otherwise, $\phi \cdot \mathcal{I}(a)$ would contain higher powers of the a_i than $\mathcal{I}(a')$. But then $\phi(\alpha)$ has to be a power of the substitution determinant. Because from

$$\begin{aligned}x_1 &= \alpha_{11}x_1' + \alpha_{12}x_2',\\ x_2 &= \alpha_{21}x_1' + \alpha_{22}x_2'\end{aligned}$$

it follows that

$$x_1' = \frac{1}{\delta}(\alpha_{22}x_1 - \alpha_{12}x_2) = \alpha_{11}'x_1 + \alpha_{12}'x_2,$$

$$x_2' = \frac{1}{\delta}(-\alpha_{21}x_1 + \alpha_{11}x_2) = \alpha_{21}'x_1 + \alpha_{22}'x_2.$$

Now, we start with the transformed form f' and transform it via these formulas. Then we must recover the original form f. Hence, we get

$$\begin{aligned}\mathcal{I}(a_0, a_1, \ldots, a_n) &= \psi(\alpha') \cdot \mathcal{I}(a_0', a_1', \ldots, a_n') \\ &= \psi(\alpha') \cdot \phi(\alpha) \cdot \mathcal{I}(a_0, a_1, \ldots, a_n),\end{aligned}$$

and therefore

$$\psi(\alpha') \cdot \phi(\alpha) = 1.$$

But

$$\psi(\alpha') = \frac{\chi(\alpha)}{\delta^N},$$

where $\chi(\alpha)$ is a polynomial function of the α. Thus, it follows that

$$\chi(\alpha) \cdot \phi(\alpha) = \delta^N.$$

But, since δ is an irreducible function of the α, both ϕ and χ must be powers of δ, up to a constant factor; therefore

$$\mathcal{I}(a_0', a_1', \ldots, a_n') = c \cdot \delta^p \mathcal{I}(a_0, a_1, \ldots, a_n).$$

On the other hand, if we apply the identity transformation

$$\begin{aligned} x_1 &= x_1', \\ x_2 &= x_2', \end{aligned}$$

then we obtain

$$\mathcal{I}(a_0, a_1, \ldots, a_n) = c \cdot \mathcal{I}(a_0, a_1, \ldots, a_n),$$

whence $c = 1$, and we recover the original definition of invariants:

$$\mathcal{I}(a_0', a_1', \ldots, a_n') = \delta^p \mathcal{I}(a_0, a_1, \ldots, a_n).$$

Lecture XIV (May 20, 1897)

Theorem *The invariants of a form of odd order must have even degree g.*

To see this, observe that $ng - 2p = 0$; therefore at least one of the integers n or g has to be even, since otherwise the weight would not be an integer, which is impossible according to the definition $p = \nu_1 + 2\nu_2 + \cdots$.

An essential distinction occurs according to whether p is even or odd. In the case of odd p one uses the terms *skew in-* and *covariants*; in the case of p even, one talks of *even in-* and *covariants.* Skew invariants are very special; the first one appears for the form of order five.

The transformation

$$x_1 = x_2', \qquad x_2 = x_1'$$

switches the coefficients a. We obtain

$$\begin{aligned} a_0' &= a_n, \\ a_1' &= a_{n-1}, \\ a_2' &= a_{n-2}, \\ &\cdots \\ a_n' &= a_0. \end{aligned}$$

If $\mathcal{I}$ is an invariant, then the equation

$$\mathcal{I}(a_n, a_{n-1}, \ldots, a_0) = (-1)^p \mathcal{I}(a_0, a_1, \ldots, a_n)$$

must always be identically satisfied. Likewise, a covariant must always satisfy the equation:

$$\mathcal{C}(a_n, a_{n-1}, \ldots, a_0; x_2, x_1) = (-1)^p \mathcal{C}(a_0, a_1, \ldots, a_n; x_1, x_2).$$

Since this equation has to be identically satisfied, it provides us with a tool to express one half of the coefficients $C_0, C_1, \ldots, C_n$ in terms of the other half. Namely, if one compares the coefficients of $x_1^i x_2^{m-i}$, then one has

$$C_i(a_n, \ldots, a_0) = C_{m-i}(a_0, \ldots, a_n) \cdot (-1)^p.$$

This makes clear the importance of the distinction of whether the weight p is even or odd. Because, from the above, we have the following:

Theorem *If one permutes the coefficients* $a_0, a_1, \ldots, a_n$ *to* $a_n, a_{n-1}, \ldots, a_0$, *then an even invariant remains unchanged, whereas an odd one changes only its sign. Furthermore, the covariant coefficient* C_i *is* *changed to C_{m-i} for an even covariant, and to $-C_{m-i}$ for a skew covariant.*

The examples considered above affirm this theorem.

I.7 The number of invariants of degree g

One of the main questions in the theory of invariants concerns the complete system of all invariants. Here, we first want to embark on the computation of the number of all invariants of a given degree g and given weight p, which also determines the order of the base form, and we want to call this number $w_n(g,p)$, or simply $w(g,p)$ or w when no ambiguity can arise. But this number only accounts for the linearly independent ones: Because if $\mathcal{I}_1, \mathcal{I}_2$ are two such invariants, then

$$\mathcal{I} = \mathcal{I}_1 + \lambda \mathcal{I}_2$$

is also an invariant of degree g and weight p; but we will not count these, only those that cannot be expressed as a linear combination of the others.

We can right away make a conjecture about the number in question. In order to obtain invariants of degree g and weight p, we need to set

$$\mathcal{I} = \sum Z a_0^{\nu_0} a_1^{\nu_1} \cdots a_n^{\nu_n} ,$$

where

$$\nu_0 + \nu_1 + \cdots + \nu_n = g,$$
$$\nu_1 + 2\nu_2 + \cdots + n\nu_n = p.$$

Here we are given n, g, and p, and from these we have to determine all possible systems of nonnegative integers $\nu_0, \nu_1, \ldots, \nu_n$. This number, that is, the number of terms of degree g and weight p, is in any case finite, call it $\omega_n(g,p)$. Hence, there are $\omega_n(g,p)$ coefficients Z in $\mathcal{I}$. Furthermore, $\mathcal{I}$ must satisfy the differential equation $\mathbf{D}\mathcal{I} = 0$, that is,

$$\mathbf{D}\mathcal{I} = \sum Z\mathbf{D} a_0^{\nu_0} a_1^{\nu_1} \cdots a_n^{\nu_n} = 0.$$

The coefficient of each term must vanish, and since there are $\omega_n(g,p-1)$ such terms—the terms are of degree g and weight $p-1$—one obtains $\omega_n(g,p-1)$ linear equations for the $\omega_n(g,p)$ quantities Z. One can therefore arbitrarily choose $\{\omega_n(g,p) - \omega_n(g,p-1)\}$ of them, and all the others are then uniquely determined. If we now arbitrarily choose $\{\omega_n(g,p) - \omega_n(g,p-1)\}$ systems of values for the first $\{\omega_n(g,p) - \omega_n(g,p-1)\}$ quantities Z, but such that there are no linear relations between them, then it is easily seen that every other arbitrary system of values for the first $\{\omega_n(g,p) - \omega_n(g,p-1)\}$ quantities Z can be expressed as a linear combination of the ones already chosen; because for $\{\omega_n(g,p) - \omega_n(g,p-1)\}$ unknown coefficients we obtain the same number of equations. We therefore have reason to conjecture that there are a total of

$$w_n(g,p) = \omega_n(g,p) - \omega_n(g,p-1)$$

linearly independent invariants.

Lecture XV (May 21, 1897)

This argument, however, makes use of hypotheses that are not at all self-evident. Firstly, there is the question of whether every invariant must contain all terms of degree g and weight p and, secondly, we cannot know whether the $\omega_n(g,p-1)$ equations for the Z are really linearly independent. It is therefore necessary to give a rigorous proof for our theorem, which we want to formulate as follows.

Theorem *The number of invariants of degree g and weight p is*

$$w_n(g,p) = \omega_n(g,p) - \omega_n(g,p-1) \tag{I}$$

for a form of order $n = \frac{2p}{g}$. Here, $\omega_n(g,p)$ is the number of terms of degree g and weight p.

Namely, let

$$\omega(g,p) = w + h$$

be the number of terms of degree g and weight p. Here $h \geq 0$, because otherwise every term of degree g and weight p could be expressed linearly in terms of the first ω of the w invariants, hence also all subsequent ones. Furthermore, for each term of a polynomial function, we call the number

$$e = ng - 2p$$

the *excess of the term*. Then all the terms of an invariant of a form of order n have excess $e = 0$. Our invariants

$$\mathcal{I}_1, \mathcal{I}_2, \ldots, \mathcal{I}_w$$

have therefore all excess $e = 0$. We now list all terms of excess $e = 2$, say

$$\mathcal{K}_1, \mathcal{K}_2, \ldots, \mathcal{K}_k,$$

and it is clear that the $\mathcal{K}$ should be of degree g, and

$$k = \omega_n(g, p-1).$$

It therefore remains to show that

$$k = h.$$

We will do this by showing that neither $k > h$ nor $k < h$ is possible.

1. If $k > h$, then one could conclude the following:
We form

$$\mathcal{I}_1, \mathcal{I}_2, \ldots, \mathcal{I}_w; \mathbf{\Delta}\mathcal{K}_1, \mathbf{\Delta}\mathcal{K}_2, \ldots, \mathbf{\Delta}\mathcal{K}_k.$$

These are all terms of degree g and weight p. Their number is $w + k > w + h = \omega_n(g, p)$; therefore, there must exist at least one relation between them:

$$c_1\mathbf{\Delta}\mathcal{K}_1 + c_2\mathbf{\Delta}\mathcal{K}_2 + \cdots + c_k\mathbf{\Delta}\mathcal{K}_k + d_1\mathcal{I}_1 + d_2\mathcal{I}_2 + \cdots + d_w\mathcal{I}_w = 0.$$

If we set

$$c_1\mathcal{K}_1 + c_2\mathcal{K}_2 + \cdots + c_k\mathcal{K}_k = \mathcal{K},$$
$$d_1\mathcal{I}_1 + d_2\mathcal{I}_2 + \cdots + d_w\mathcal{I}_w = -\mathcal{I},$$

then we must have

$$\mathbf{\Delta}\mathcal{K} = \mathcal{I},$$

where $\mathcal{I}$ is an invariant. But this is impossible, which can be seen as follows. The sequence

$$\mathcal{K}, \mathbf{D}\mathcal{K}, \mathbf{D}^2\mathcal{K}, \ldots, \mathbf{D}^{l-1}\mathcal{K}, \ \mathbf{D}^l\mathcal{K}, \ldots$$

has to terminate, say at $\mathbf{D}^{l-1}\mathcal{K} \neq 0$. Now we have

$$(\mathbf{D}^l\mathbf{\Delta} - \mathbf{\Delta}\mathbf{D}^l)\mathcal{K} = l(ng - 2p + l + 1)\mathbf{D}^{l-1}\mathcal{K}.$$

Certainly, $l \neq 0$, otherwise $\mathcal{K}$ would be 0, which is not possible since the $\mathcal{K}_i$ all have distinct terms. Therefore $\mathbf{D}^l\mathbf{\Delta}\mathcal{K} = \mathbf{D}^l\mathcal{I} = 0$, and it now follows that

$$l = -1,$$

which is impossible. Hence, the assumption that $k > h$ cannot be correct.

2. $k < h$ cannot hold either. Because if we attempt to list all possible invariants, then we obtain k' linearly independent equations for the $w + h$ unknown quantities Z, where $k' \leq k < h$. Then there would be

$w+h-k'$ linearly independent invariants, which is more than w, so we obtain a contradiction to our assumption.

Therefore, $k=h$, and our theorem is proven.

Our previous examples can be used for verification. We saw (Section I.6) that there is only one invariant for $n=4$, $g=2$, $p=4$, and if we list all the terms for $g=2$, $p=4$ as well as those for $g=2$, $p=3$, then we obtain

$$a_0a_4, \quad a_1a_3, \quad a_2^2 \qquad \omega_4(2,4)=3;$$
$$a_0a_3, \quad a_1a_2 \qquad \omega_4(2,3)=2;$$

hence

$$w_4(2,4)=1.$$

Likewise, for $n=4$, $g=3$, $p=6$ one has

$$g=3,\ p=6: \quad a_0a_2a_4,\ a_0a_3^2,\ a_1a_2a_3,\ a_2^3,\ a_1^2a_4 \qquad \omega_4(3,6)=5;$$
$$g=3,\ p-1=5: \quad a_0a_1a_4,\ a_0a_2a_3,\ a_1^2a_3,\ a_1a_2^2 \qquad \omega_4(3,5)=4;$$

hence

$$w_4(3,6)=1.$$

The formula we have proved is of fundamental importance. One sees, however, that it would be very difficult and tedious to derive the number of invariants in each individual case with its help. We can, however, easily derive from it a much simpler formula for w, which then makes the calculation of w very easy.

Lecture XVI (May 31, 1897)

The number $\omega_n(g,p)$ admits the following representation:

$$\omega_n(g,p) = \left[\frac{1}{(1-y)(1-xy)(1-x^2y)\cdots(1-x^ny)}\right]_{x^py^g}. \tag{1}$$

This notation means that if one expresses the function $\frac{1}{(1-y)\cdots(1-x^ny)}$ in terms of powers of x and y, then the coefficient of x^py^g is equal to $\omega_n(g,p)$. This is because

$$\frac{1}{(1-y)(1-xy)\cdots(1-x^ny)} = (1+y+y^2+\cdots)(1+xy+x^2y^2+\cdots)$$
$$\times(1+x^2y+x^4y^2+\cdots)\cdots(1+x^ny+x^{2n}y^2+\cdots),$$

so that the general term is:

$$y^{\nu_0} x^{\nu_1} y^{\nu_1} x^{2\nu_2} y^{\nu_2} \cdots x^{n\nu_n} y^{\nu_n} = x^{\nu_1+2\nu_2+\cdots+n\nu_n}\, y^{\nu_0+\nu_1+\cdots+\nu_n}.$$

The coefficient of the term $x^p y^g$ will be such that p can be expressed in the form $\nu_1 + 2\nu_2 + \cdots + n\nu_n$ and, simultaneously, g can be expressed in the form $\nu_0 + \nu_1 + \cdots + \nu_n$, and hence will be equal to $\omega_n(g,p)$, as asserted.

Formula (1) can always be written in such a way that it contains only one variable. If we expand the fraction in question in terms of powers of y, then we get

$$\frac{1}{(1-y)(1-xy)\cdots(1-x^n y)} = 1 + C_1 y + C_2 y^2 + C_3 y^3 + \cdots,$$

where the C_i are functions of the x only. Furthermore,

$$(1-y)\,\frac{1}{(1-y)(1-xy)\cdots(1-x^n y)} = \frac{1-x^{n+1}y}{(1-xy)(1-x^2 y)\cdots(1-x^n y)(1-x^{n+1}y)};$$

hence

$$(1-y)(1 + C_1 y + C_2 y^2 + \cdots) = (1-x^{n+1}y)(1 + C_1 xy + C_2 x^2 y^2 + \cdots).$$

If we determine the coefficient of y^g on both sides, then we get

$$C_g - C_{g-1} = C_g x^g - C_{g-1} x^{g+n};$$

hence

$$C_g = \frac{1-x^{n+g}}{1-x^g}\, C_{g-1}.$$

Successive application of this formula for $g, g-1, \ldots, 2, 1$ and subsequent multiplication of all formulas results in:

$$C_g = \frac{(1-x^{n+g})(1-x^{n+g-1})\cdots(1-x^{n+2})(1-x^{n+1})}{(1-x^g)(1-x^{g-1})\cdots(1-x^2)(1-x)}.$$

The coefficient of x^p in this expansion is again $\omega_n(g,p)$, that is,

$$\omega_n(g,p) = \left(\frac{(1-x^{n+1})(1-x^{n+2})\cdots(1-x^{n+g})}{(1-x)(1-x^2)\cdots(1-x^g)}\right)_{x^p}. \tag{2}$$

So, if we set

$$\frac{(1-x^{n+1})(1-x^{n+2})\cdots(1-x^{n+g})}{(1-x)(1-x^2)\cdots(1-x^g)} = c_0 + c_1 x + c_2 x^2 + \cdots + c_p x^p + \cdots,$$

then

$$c_p = \omega_n(g,p)$$

and therefore

$$w_n(g,p) = c_p - c_{p-1}.$$

But we also have

$$\begin{aligned}(1-x)\,&\frac{(1-x^{n+1})(1-x^{n+2})\cdots(1-x^{n+g})}{(1-x)(1-x^2)\cdots(1-x^g)}\\ &= (1-x)(c_0+c_1x+\cdots+c_px^p+\cdots)\\ &= c_0+x(c_1-c_0)+\cdots+x^p(c_p-c_{p-1})+\cdots,\end{aligned}$$

so $w_n(g,p)$ is again the coefficient of x^p in this last expansion. And so we have the important and easily usable formula:

$$w_n(g,p) = \left\{\frac{(1-x^{n+1})(1-x^{n+2})\cdots(1-x^{n+g})}{(1-x^2)(1-x^3)\cdots(1-x^g)}\right\}_{x^p}. \qquad (II)$$

We immediately use this formula to derive the "full invariant systems" for forms of order two, three, four, that is, to list all possible invariants from which all others can be derived as polynomial expressions, but such that none of them is a polynomial function of the others.

1. $n=2$. Here $p=g$. We therefore have

$$w_2(g,g) = \left\{\frac{(1-x^3)(1-x^4)\cdots(1-x^{g+2})}{(1-x^2)(1-x^3)\cdots(1-x^g)}\right\}_{x^g},$$

or

$$w_2(g,g) = \left\{\frac{(1-x^{g+1})(1-x^{g+2})}{1-x^2}\right\}_{x^g}.$$

But if one expands $\frac{1}{1-x^2}$, and then multiplies by the numerator, then all terms of the numerator except 1 result in summands of degree higher than g, and so can be omitted. Therefore,

$$w_2(g,g) = \left(\frac{1}{1-x^2}\right)_{x^g},$$

so

$$\begin{aligned} w_2(g,g) &= 0 \qquad \text{if } g \text{ is odd},\\ w_2(g,g) &= 1 \qquad \text{if } g \text{ is even}.\end{aligned}$$

Thus, there are no invariants of odd degree. There is one invariant of degree two, namely,

$$\mathcal{I} = a_0 a_2 - a_1^2,$$

from which it follows that $\mathcal{I}^r$ is also an invariant, and is of even degree. But since there is only one invariant of degree $2r$, we have the following result.

Theorem *There is only one invariant of a quadratic form, namely,* $\mathcal{I} = a_0 a_2 - a_1^2$.

Note that we omit those invariants that are polynomial functions of $\mathcal{I}$.

Lecture XVII (June 1, 1897)

2. $n = 3$. Then $p = \frac{3g}{2}$. We have

$$w\left(g, \frac{3g}{2}\right) = \left\{ \frac{(1-x^4)(1-x^5)\cdots(1-x^{g+3})}{(1-x^2)(1-x^3)\cdots(1-x^g)} \right\}_{x^{\frac{3g}{2}}}$$

$$= \left\{ \frac{(1-x^{g+1})(1-x^{g+2})(1-x^{g+3})}{(1-x^2)(1-x^3)} \right\}_{x^{\frac{3g}{2}}}$$

or, since we can omit a number of terms,

$$w\left(g, \frac{3g}{2}\right) = \left\{ \frac{1 - x^{g+1} - x^{g+2} - x^{g+3}}{(1-x^2)(1-x^3)} \right\}_{x^{\frac{3g}{2}}}$$

$$= \left\{ \frac{1}{(1-x^2)(1-x^3)} \right\}_{x^{\frac{3g}{2}}} - \left\{ \frac{x^{g+1}(1+x+x^2)}{(1-x^2)(1-x^3)} \right\}_{x^{\frac{3g}{2}}}$$

$$= \left\{ \frac{1}{(1-x^2)(1-x^3)} \right\}_{x^{\frac{3g}{2}}} - \left\{ \frac{x^{g+1}}{(1-x)(1-x^2)} \right\}_{x^{\frac{3g}{2}}}$$

$$= \left\{ \frac{1}{(1-x^2)(1-x^3)} \right\}_{x^{\frac{3g}{2}}} - \left\{ \frac{x}{(1-x)(1-x^2)} \right\}_{x^{\frac{g}{2}}},$$

since it is clearly irrelevant whether we determine the coefficient of x^{i+k} in $x^i\mathcal{P}(x)$ or the coefficient of x^k in $\mathcal{P}(x)$. Likewise, it is irrelevant whether we determine the coefficient of x^k in $\mathcal{P}(x)$ or that of x^{2k} in

$\mathcal{P}(x^2)$. Therefore, we further have:

$$w\left(g, \frac{3g}{2}\right) = \left\{\frac{1}{(1-x^4)(1-x^6)}\right\}_{x^{3g}} - \left\{\frac{x^2}{(1-x^2)(1-x^4)}\right\}_{x^g}$$

$$= \left\{\frac{1}{(1-x^4)(1-x^6)}\right\}_{x^{3g}}$$

$$+ \left\{\frac{1}{1-x^4}\right\}_{x^g} - \left\{\frac{1}{(1-x^2)(1-x^4)}\right\}_{x^g}$$

$$= \left\{\frac{1}{1-x^4}\right\}_{x^g} + \left\{\frac{1}{(1-x^4)(1-x^6)} - \frac{1}{(1-x^6)(1-x^{12})}\right\}_{x^{3g}}$$

$$= \left\{\frac{1}{1-x^4}\right\}_{x^g} + \left\{\frac{1+x^4+x^8-1}{(1-x^6)(1-x^{12})}\right\}_{x^{3g}}.$$

Calculation shows that the second part contains only terms of the form x^{4m+3n}, where $m = 1$ or 2, hence none of the form x^{3g}, and so does not contribute anything. And thus it follows that

$$w\left(g, \frac{3g}{2}\right) = \left\{\frac{1}{1-x^4}\right\}_{x^g}.$$

From this it follows readily that there are no invariants if g is not divisible by 4, one invariant if g is divisible by 4. One therefore has the following:

Theorem *A cubic form has only one invariant, namely,*

$$\mathcal{I} = a_0^2\, a_3^2 - 3\, a_1^2\, a_2^2 + 4a_1^3\, a_3 + 4a_0\, a_2^3 - 6a_0\, a_1\, a_2\, a_3,$$

the so-called discriminant of the cubic form.

3. $n = 4$. Hence $p = 2g$. We have

$$w(g, 2g) = \left\{\frac{(1-x^{g+1})(1-x^{g+2})(1-x^{g+3})(1-x^{g+4})}{(1-x^2)(1-x^3)(1-x^4)}\right\}_{x^{2g}}$$

$$= \left\{\frac{1-x^{g+1}-x^{g+2}-x^{g+3}-x^{g+4}}{(1-x^2)(1-x^3)(1-x^4)}\right\}_{x^{2g}}.$$

The following calculation is also based on the same principles as in the previous case. One has:

$w(g, 2g)$

$$= \left\{\frac{1}{(1-x^2)(1-x^3)(1-x^4)}\right\}_{x^{2g}} - \left\{\frac{x^{g+1}(1+x+x^2+x^3)}{(1-x^2)(1-x^3)(1-x^4)}\right\}_{x^{2g}}$$

or

$$
\begin{aligned}
w(g, 2g) & \\
&= \left\{\frac{1}{(1-x^2)(1-x^3)(1-x^4)}\right\}_{x^{2g}} - \left\{\frac{x}{(1-x)(1-x^2)(1-x^3)}\right\}_{x^g} \\
&= \left\{\frac{1}{(1-x^2)(1-x^3)(1-x^4)}\right\}_{x^{2g}} - \left\{\frac{x^2}{(1-x^2)(1-x^4)(1-x^6)}\right\}_{x^{2g}} \\
&= \left\{\frac{1+x^3-x^2}{(1-x^2)(1-x^4)(1-x^6)}\right\}_{x^{2g}} \\
&= \left\{\frac{1-x^2}{(1-x^2)(1-x^4)(1-x^6)}\right\}_{x^{2g}} + \qquad 0 \\
&= \left\{\frac{1}{(1-x^4)(1-x^6)}\right\}_{x^{2g}},
\end{aligned}
$$

or, finally,

$$w(g, 2g) = \left\{\frac{1}{(1-x^2)(1-x^3)}\right\}_{x^g}.$$

Lecture XVIII (June 3, 1897)

Calculation gives

$$w(g, 2g) = \left\{(1 + x^2 + x^4 + \cdots)(1 + x^3 + x^6 + \cdots)\right\}_{x^g};$$

hence there are as many independent invariants of degree g as there are positive integer solutions of the equation

$$2k + 3l = g.$$

Hence, for $g = 2$ there is one invariant, just as for $g = 3$, and these are (cf. Section I.6):

$$
\begin{aligned}
\mathcal{I}_2 &= a_0 a_4 - 4a_1 a_3 + 3a_2^2, \\
\mathcal{I}_3 &= a_0 a_2 a_4 - a_0 a_3^2 - a_1^2 a_4 + 2a_1 a_2 a_3 - a_2^3.
\end{aligned}
$$

But then

$$\mathcal{I}_2^k \mathcal{I}_3^l = \mathcal{I}$$

is also an invariant; that is, one can form as many invariants of degree g from $\mathcal{I}_2$ and $\mathcal{I}_3$ as there are nonnegative integer solutions of the equation

$$2k + 3l = g.$$

Furthermore, these are all linearly independent. Because, if there existed a linear relation

$$\sum c_{kl} \mathcal{I}_2^k \mathcal{I}_3^l = 0$$

between them, then it would have to be valid identically for all a; and so, for instance, for $a_1 = 0$, $a_2 = 0$. Thus, we would also have

$$\sum c_{kl}\, a_0^{k+l}\, a_4^k\, a_3^{2l}\, (-1)^l = 0$$

identically. But all k, as well as all l, are distinct, since every k determines the corresponding l and vice versa. Thus, all the terms in the sum are distinct, which shows that such a relation is not possible. From this follows immediately the theorem:

Theorem *The full invariant system of a binary form of order four is formed by the two invariants $\mathcal{I}_2$, $\mathcal{I}_3$.*

It would lead too far to carry out a general analysis of more cases, especially since the calculation becomes increasingly difficult. The form of order five is of particular interest, because it is here that the first skew invariant appears. Sylvester has studied this case extensively. We mention the results of interest to us. For $n = 5$ we have

$$\begin{aligned} w\left(g, \frac{5g}{2}\right) &= \left\{\frac{1 - x^{36}}{(1-x^4)(1-x^8)(1-x^{12})(1-x^{18})}\right\}_{x^g} \\ &= \{(1 + x^4 + x^8 + \cdots)(1 + x^8 + x^{16} + \cdots) \\ &\qquad (1 + x^{12} + x^{24} + \cdots)(1 + x^{18} + x^{36} + \cdots)(1 - x^{36})\}_{x^g} . \end{aligned}$$

There are no invariants of odd degree. The invariant of lowest degree is the one of degree four. If one counts how many linearly independent invariants of degree g can be formed from those of lower degrees and, on the other hand, how many our last formula predicts, then one finds the following:

$g = 4$	1 inv.	1 inv. $\mathcal{I}_4$	deg. $g = 4$	weight $p = 10$
$g = 8$	2 inv.	1 new inv. $\mathcal{I}_8$	" $g = 8$	" $p = 20$
$g = 12$	3 inv.	1 new inv. $\mathcal{I}_{12}$	" $g = 12$	" $p = 30$
$g = 16$	3 inv.			
$g = 18$	1 inv.	1 new inv. $\mathcal{I}_{18}$	" $g = 18$	" $p = 45$.

For $g = 36$, one finds furthermore that one can construct one more invariant from $\mathcal{I}_4$, $\mathcal{I}_8$, $\mathcal{I}_{12}$, $\mathcal{I}_{18}$ than the formula for $g = 36$ predicts. Hence there must exist a linear relation between them. Thus, there must exist a relation between $\mathcal{I}_4$, $\mathcal{I}_8$, $\mathcal{I}_{12}$, $\mathcal{I}_{18}$, a so-called *syzygy*, which is of degree 36 in the a. Aside from this relation, there are no more special phenomena to observe.

Theorem *The full invariant system of the form of order five is given by the four invariants $\mathcal{I}_4$, $\mathcal{I}_8$, $\mathcal{I}_{12}$, $\mathcal{I}_{18}$, between which there exists a syzygy of degree 36.*

Sylvester has calculated these invariants and the syzygy and found that, if one expresses the invariants in a form which has a minimal number of terms, then these numbers are as follows:

$\mathcal{I}_4$	12 terms
$\mathcal{I}_8$	59 terms
$\mathcal{I}_{12}$	228 terms
$\mathcal{I}_{18}$	848 terms

The latter have, for the most part, very large coefficients (listed in Faà di Bruno (1876) and Walter (1881)).

For $n = 6$ we obtain results similar to $n = 5$.

We end this section with a general theorem discovered by Hermite, the so-called Reciprocity Law. We found (Lecture XII) that the number of terms of degree g and weight $p = \frac{ng}{2}$ has to be

$$\omega_n\left(g, \frac{ng}{2}\right) = \left\{\frac{(1 - x^{n+1})(1 - x^{n+2})(1 - x^{n+3})\cdots(1 - x^{n+g})}{(1 - x)(1 - x^2)\cdots(1 - x^g)}\right\}_{x^{\frac{ng}{2}}},$$

which we can also write as

$$\omega_n\left(g, \frac{ng}{2}\right) = \left\{\frac{(1 - x^{g+1})(1 - x^{g+2})\cdots(1 - x^{g+n})}{(1 - x)(1 - x^2)\cdots(1 - x^n)}\right\}_{x^{\frac{ng}{2}}}, \tag{3}$$

regardless of whether $g \geq n$ or not. If we interchange n and g, then the number of terms of degree n and weight $\frac{ng}{2}$ is, according to the first formula,

$$\omega_g\left(n, \frac{ng}{2}\right) = \left\{ \frac{(1-x^{g+1})(1-x^{g+2})\cdots(1-x^{g+n})}{(1-x)(1-x^2)\cdots(1-x^n)} \right\}_{x^{\frac{ng}{2}}}.$$

We therefore have that

$$\omega_n\left(g, \frac{ng}{2}\right) = \omega_g\left(n, \frac{ng}{2}\right), \tag{4}$$

and, since

$$w_n\left(g, \frac{ng}{2}\right) = \omega_n\left(g, \frac{ng}{2}\right) - \omega_n\left(g, \frac{ng}{2} - 1\right),$$

it follows that

$$w_n\left(g, \frac{ng}{2}\right) = w_g\left(n, \frac{ng}{2}\right). \tag{III}$$

This is the Reciprocity Law.

Reciprocity Law of Hermite *The number of invariants of a base form of order n and degree g is equal to the number of invariants of a base form of order g and degree n.*

In particular, we have the following.

Theorem *A linear form has no invariants; likewise, there is no base form with an invariant that is linear in the coefficients. A base form of even order has a quadratic invariant; a form of odd order does not. A form of order n has a cubic invariant if and only if n is divisible by four. And so forth.*

One can formulate the reciprocity law also in a different, number-theoretic form. Namely, $\omega_n(g, p)$ is equal to the number of representations of p as a sum of g integers from the sequence 0, 1, 2, ..., n. This can be seen as follows. A term

$$a_0^{\nu_0} a_1^{\nu_1} \cdots a_n^{\nu_n}$$

of degree g and weight p determines such a representation of p, namely,

$$\underbrace{(0+0+\cdots+0)}_{\nu_0 \text{ times}} + \underbrace{(1+1+\cdots+1)}_{\nu_1 \text{ times}} + \cdots + \underbrace{(n+n+\cdots+n)}_{\nu_n \text{ times}} = p,$$

where the number of summands is g. Conversely, given such an arbitrary representation, one can immediately form a term of degree g and weight p. If we fix the way one is derived from the other, then the relationship is a one-to-one correspondence, which proves the assertion.

Since we have in general, for arbitrary p,

$$\omega_n(g,p) = \omega_g(n,p), \tag{4'}$$

this implies the following:

Theorem *Any integer p can be written as a sum of n numbers from the sequence 0, 1, 2, ..., g in as many ways as it can be written as a sum of g numbers from the sequence 0, 1, 2, ..., n.*

The counting method of Sylvester, which we have described in this section and which can be adapted to the case of covariants, is apparently a very clever and often convenient method, but we can nevertheless not be satisfied with merely knowing the number of invariants, as it is even more important to also know about the in- and covariants themselves, and about the relations between them. The Sylvester counting method does not produce general results of fundamental importance.

We therefore need to search for new methods. In the following section we want to show that every in- and covariant of a form can be expressed as a polynomial function of the in- and covariants of degrees two and three—aside from the base form itself.* Therefore, their determination will occupy us at the beginning of the following section.

Lecture XIX (June 4, 1897)

I.8 The invariants and covariants of degrees two and three

If one considers the problem of determining all possible in- and covariants of low degrees, one sees easily that for degree $g = 1$ there is only one covariant, namely, the form f itself. This is because $C_0 = a_p$ satisfies the differential equation $\mathbf{D}C_0 = 0$ only if $a_p = a_0$.

We therefore move on immediately to the determination of the *covariants of degree two*. A covariant is completely determined by its source, so we restrict ourselves to determining those. There are no covariants of degree two with odd weight, which we will not prove. If the weight is an even number,

$$p = 2\pi,$$

* That is, as a quotient of a polynomial in the numerator and a power of the base form in the denominator.

then there is exactly one covariant associated to each p between 2 and n, respectively $n-1$, namely, the so-called *pth transvection* of the form f over itself, which we will not prove either. This covariant is

$$f_p = \left\{ a_0 a_p - \binom{p}{1} a_1 a_{p-1} + \binom{p}{2} a_2 a_{p-2} - \cdots \right.$$

$$+ (-1)^{\pi-1} \binom{p}{\pi-1} a_{\pi-1} a_{\pi+1}$$

$$\left. + (-1)^{\pi} \frac{1}{2} \binom{p}{\pi} a_\pi^2 \right\} x_1^{2n-2p} + \cdots .$$

The covariant property of this function can be proved using the principles of Section I.6. The coefficient C_0 of x_1^{2n-2p} is a homogeneous isobaric function of degree two and weight p, and the order $m = 2n - 2p$ of the covariant satisfies the condition $m = ng - 2p$. Furthermore, C_0 satisfies the differential equation $\mathbf{D}C_0 = 0$, which can perhaps most easily be seen by putting C_0 in the form

$$C_0 = \sum_{i=0}^{\pi-1} \left\{ (-1)^i a_i a_{p-i} \binom{p}{i} \right\} + (-1)^{\pi} \frac{1}{2} \binom{p}{\pi} a_\pi^2 ,$$

and then applying the operator $\mathbf{D}$ to the summands. One can then deduce the property $\mathbf{D}C_0 = 0$ easily by rewriting the resulting expressions appropriately (separation of one term from the sum, interchanging i and $i-1$, etc.). This proves the covariant property.

Therefore, there is one covariant of degree two for each even weight from $p = 2$ to $p = n$, respectively $n-1$. If the base form has even order, then there is an invariant of degree 2 for $p = n$, and only one according to the Reciprocity Law. For $p = 2, 4, 6, \ldots$, the coefficients of x_1^{2n-2p} are

$$\begin{aligned} c(f_2) &= a_0 a_2 - a_1^2 = c_2, \\ c(f_4) &= a_0 a_4 - 4 a_1 a_3 + 3 a_2^2 = c_4, \\ c(f_6) &= a_0 a_6 - 6 a_1 a_5 + 15 a_2 a_4 - 10 a_3^2 = c_6, \\ & \cdots . \end{aligned}$$

If one considers the forms of order two, four, six, ..., respectively, then these are the quadratic invariants of the forms of order two, four, six, ..., respectively.

Regarding the *covariants of degree three*, they all have odd weight

$$p = 2\pi + 1$$

and are those which occur in the following expression, where $p = 3, 5, 7, \dots, n$, respectively $n-1$:

$$f_p = \Big\{ a_0 \Big[a_0 a_p - (p-2)a_1 a_{p-1} + \frac{p-1}{1\cdot 2}(p-4)a_2 a_{p-2}$$

$$- \frac{(p-1)\,(p-2)}{1\cdot 2\cdot 3}(p-6)a_3 a_{p-3} + \cdots + (-1)^\pi \binom{p-1}{\pi-1}\frac{1}{\pi}a_\pi a_{\pi+1}\Big]$$

$$- 2a_1 \Big[a_0 a_{p-1} - \binom{p-1}{1} a_1 a_{p-2} + \binom{p-1}{2} a_2 a_{p-3} - \cdots$$

$$+ (-1)^{\pi-1}\binom{p-1}{\pi-1} a_{\pi-1} a_{\pi+1}$$

$$+ (-1)^\pi \frac{1}{2}\binom{p-1}{\pi} a_\pi^2 \Big]\Big\} x_1^{3n-2p} + \cdots .$$

Homogeneity and isobarity of the source c_p are apparent; the condition $m = ng - 2p$ is satisfied as well. That $\mathbf{D}c_p = 0$ can be seen most easily as follows. If one sets

$$A = a_0 a_{p-1} - \binom{p-1}{1} a_1 a_{p-2} + \cdots + (-1)^\pi \frac{1}{2}\binom{p-1}{\pi} a_\pi^2,$$

then we have from above that

$$\mathbf{D}A = 0,$$

since $p-1$ is even, and, furthermore, one sees easily that

$$a_0 a_p - (p-2)a_1 a_{p-1} + \cdots = \frac{1}{n-p+1}\mathbf{\Delta}A .$$

Therefore

$$c_p = \frac{a_0}{n-p+1}\mathbf{\Delta}A - 2a_1 A,$$

from which it follows that

$$\mathbf{D}c_p = \frac{a_o}{n-p+1}\mathbf{D\Delta}A - 2a_1\mathbf{D}A - 2Aa_0$$

$$= \frac{1}{n-p+1}\left\{ a_0\big(\mathbf{D\Delta}A - 2(n-p+1)A\big)\right\}.$$

Since (Section I.5)

$$(\mathbf{D\Delta} - \mathbf{\Delta D})A = \big(2n - 2(p-1)\big)A,$$

it follows that

$$\mathbf{D}\mathbf{\Delta}A = 2(n-p+1)A,$$

so that, indeed,

$$\mathbf{D}c_p = 0.$$

For $p = 3$ one obtains, for example,

$$\begin{aligned} c_3 &= a_0[a_0a_3 - a_1a_2] - 2a_1[a_0a_2 - a_1^2] \\ &= a_0^2a_3 - 3a_0a_1a_2 + 2a_1^3. \end{aligned}$$

If we now add the covariants $f \cdot f_p$, where f_p runs through the covariants of degree two for even p, to the covariants f_p of degree three for odd p, then we have the complete in- and covariant system of degree three. Of course, we again omit the proof that there are no others.

Lecture XX (June 4, 1897)

The covariants f_p are of great importance, even more so since we have the following:

Theorem *Every covariant of the base form f can be expressed as a polynomial function of the covariants*

$$f_1,\ f_2,\ f_3,\ \ldots,\ f_n,$$

up to a power of the base form itself, appearing in the denominator. We obtain the representation if we substitute

$$\begin{aligned} a_0 &= f, \\ a_1 &= 0 \end{aligned}$$

in the source of the covariant, and if for the remaining coefficients

$$a_n,\ a_{n-1},\ \ldots,\ a_2$$

we substitute

$$\begin{aligned} a_p = \frac{1}{f}\Bigg\{ f_p - \binom{p}{2} a_2a_{p-2} + \cdots \\ +(-1)^\pi \binom{p}{\pi-1} a_{\pi-1}a_{\pi+1} + (-1)^{\pi+1}\frac{1}{2}\binom{p}{\pi} a_\pi^2 \Bigg\}, \end{aligned}$$

if $p = 2\pi$ *is even, or*

$$a_p = \frac{f_p}{f^2} + \frac{1}{f}\Big[-\binom{p-1}{1}\frac{1}{2}(p-4)a_2a_{p-2} + \cdots$$
$$+(-1)^{\pi+1}\binom{p-1}{\pi-1}\frac{1}{\pi}a_\pi a_{\pi+1}\Big],$$

if $p = 2\pi + 1$ *is odd.*

That is, if C is a covariant of the form f, then one can write

$$C = \frac{G(f_1, f_2, f_3, \ldots, f_n)}{f^N},$$

where G is a polynomial function of its arguments.

We proceed to the proof of the theorem. Let

$$C = C_0 x_1^m + \cdots$$

be the covariant, so $C_0 = C_0(a_0, a_1, \ldots, a_n)$ is its source, which therefore satisfies the usual conditions. We now substitute into C_0 the sources of the $n-1$ covariants

$$f_2,\ f_3,\ \ldots,\ f_n$$

for

$$a_2,\ a_3,\ \ldots, a_n,$$

that is, the quantities

$$c_2,\ c_3,\ \ldots, c_n.$$

This can easily be done since one sees immediately from the formulas for c_p that

$$a_p = \frac{1}{a_0^\rho}\, g(a_0,\ a_1, \ldots, a_{p-1};\ c_p),$$

where ρ is either 1 or 2 and g is a polynomial function. One will therefore simply express a_n through $a_0, a_1, \ldots, a_{n-1}, c_n$, and subsequently a_{n-1} through $a_0,\ a_1,\ \ldots,\ a_{n-2},\ c_n$, etc. The occurring functions are always polynomial functions up to a power of a_0. If one replaces f by a_0 and f_p by c_p in the formulas given in the statement of our theorem, then one obtains the formulas that express a_p through $a_0,\ a_1,\ \ldots,\ a_{p-1},\ c_p$, up to terms which contain a_1. From all this one sees that

$$C_0 = \frac{\Gamma_0(a_0, a_1;\ c_2, c_3, \ldots, c_n)}{a_0^N},$$

where Γ_0 is again a polynomial function. Obviously, we now have:

$$\frac{\partial \Gamma_0}{\partial a_i}(a) = \frac{\partial \Gamma_0}{\partial c_2}\frac{\partial c_2}{\partial a_i} + \frac{\partial \Gamma_0}{\partial c_3}\frac{\partial c_3}{\partial a_i} + \cdots + \frac{\partial \Gamma_0}{\partial c_n}\frac{\partial c_n}{\partial a_i}, \qquad i = 2, \ldots, n,$$

and

$$\frac{\partial \Gamma_0}{\partial a_1}(a) = \frac{\partial \Gamma_0}{\partial a_1} + \frac{\partial \Gamma_0}{\partial c_2}\frac{\partial c_2}{\partial a_1} + \cdots + \frac{\partial \Gamma_0}{\partial c_n}\frac{\partial c_n}{\partial a_1}.$$

But since C_0 satisfies the differential equation $\mathbf{D}C_0 = 0$, it follows that

$$\mathbf{D}\Gamma_0 = 0$$

also holds; explicitly:

$$a_0\frac{\partial \Gamma_0}{\partial a_1} + \frac{\partial \Gamma_0}{\partial c_2}\mathbf{D}c_2 + \frac{\partial \Gamma_0}{\partial c_3}\mathbf{D}c_3 + \cdots + \frac{\partial \Gamma_0}{\partial c_n}\mathbf{D}c_n = 0.$$

Since, on the other hand, $c_2, c_3, \ldots, c_n$ are covariant sources themselves, it follows that $\mathbf{D}c_2 = 0$, $\mathbf{D}c_3 = 0$, ..., $\mathbf{D}c_n = 0$, so we further have

$$\frac{\partial \Gamma_0}{\partial a_1} = 0;$$

that is, if one substitutes $c_2, c_3, \ldots, c_n, a_0, a_1$ for $a_0, a_1, \ldots, a_n$ in C_0, then the resulting expession does not contain a_1 anymore. We therefore have

$$a_0^N C_0 = \Gamma_0(a_0; c_2, c_3, \ldots, c_n).$$

Here, all the terms in Γ_0 have the same weight, since this is the case in C_0, and substitution of the c_i for the a_i does not change the weight, according to the formulas. If one therefore replaces a_0 by f, c_2 by f_2, c_3 by f_3, ..., c_n by f_n in Γ_0, then all summands are changed by the same power of the determinant after applying a linear transformation. Thus $\Gamma_0(f; f_2, f_3, \ldots, f_n)$ is a covariant. But then it follows that

$$f^N C_0 = \Gamma_0(f; f_2, f_3, \ldots, f_n),$$

because both sides of the equation are covariants with the same source. They are therefore identical if one can prove in addition that Γ_0 is homogeneous in the x. But this can be seen as follows. Let

$$a_0^{\rho_0} c_2^{\rho_2} c_3^{\rho_3} \cdots c_n^{\rho_n},$$

respectively

$$f^{\rho_0} f_2^{\rho_2} f_3^{\rho_3} \cdots f_n^{\rho_n},$$

be an arbitrary term in the expansion of Γ_0. The order of this term with respect to the x is

$$\begin{aligned} n\rho_0 &+ (2n-4)\rho_2 + (2n-8)\rho_4 + \cdots \\ &+ \big(2n-(2n-4)\big)\rho_{n-2} + (2n-2n)\rho_n \\ &+ (3n-6)\rho_3 + (3n-10)\rho_5 + \cdots \\ &+ \big(3n-(2n-2)\big)\rho_{n-1} \end{aligned}$$

if n is even, and

$$\begin{aligned} n\rho_0 &+ (2n-4)\rho_2 + (2n-8)\rho_4 + \cdots + \big(2n-(2n-2)\big)\rho_{n-1} \\ &+ (3n-6)\rho_3 + (3n-10)\rho_5 + \cdots + (3n-2n)\rho_n \end{aligned}$$

if n is odd. And since the weight of the term $f^{\rho_0} f_2^{\rho_2} \cdots f_n^{\rho_n}$ is equal to the weight p of C_0, we have

$$2\rho_2 + 3\rho_3 + \cdots + n\rho_n = p.$$

It now follows easily that this order then is

$$\begin{aligned} &n\,\{\rho_0 + 2(\rho_2+\rho_4+\cdots+\rho_n) \\ &\qquad + 3(\rho_3+\rho_5+\cdots+\rho_{n-1})\} - 2p \qquad \text{for } n \text{ even,} \\ &n\,\{\rho_0 + 2(\rho_2+\rho_4+\cdots+\rho_{n-1}) \\ &\qquad + 3(\rho_3+\rho_5+\cdots+\rho_n)\} - 2p \qquad \text{for } n \text{ odd.} \end{aligned}$$

If we now express

$$a_0^{\rho_0} c_2^{\rho_2} \cdots c_n^{\rho_n}$$

in terms of the a, then it has to be homogeneous of degree $N+g$, where g is the degree of C_0, because $a_0^N C_0 = \Gamma_0$. This is because, if there were terms of higher or lower degree, then they would have to cancel out, and would therefore cause all terms appearing in $a_0^{\rho_0} c_2^{\rho_2} \cdots c_n^{\rho_n}$ to cancel each other out, so that $a_0^{\rho_0} c_2^{\rho_2} \cdots c_n^{\rho_n}$ would vanish and could therefore be ignored. Thus we have

$$\begin{aligned} &\rho_0 + 2(\rho_2+\rho_4+\cdots+\rho_n) + 3(\rho_3+\rho_5+\cdots+\rho_{n-1}) \\ &\qquad = N+g \qquad \text{if } n \text{ is even,} \\ &\rho_0 + 2(\rho_2+\rho_4+\cdots+\rho_{n-1}) + 3(\rho_3+\rho_5+\cdots+\rho_n) \\ &\qquad = N+g \qquad \text{if } n \text{ is odd.} \end{aligned}$$

But this implies that the desired order of the term $f^{\rho_0} f_2^{\rho_2} \cdots$ is

$$n(N+g) - 2p = nN + ng - 2p = nN + m.$$

It is independent of the term we consider, and is equal to the order of $f^N\mathcal{C}$. Herewith our theorem is completely proved. Since Γ_0 does not contain a_1 anymore, we can set $a_1 = 0$ in the beginning.

We right away give applications of our theorem.

I. $n = 2$. An in- or covariant of f must be a rational function of

$$f = a_0x_1^2 + 2a_1x_1x_2 + a_2x_2^2 = f_{1,2},$$
$$f_2 = a_0a_2 - a_1^2 = \mathcal{H}_{2,0}$$

of the form

$$\mathcal{C} = \frac{f^M f_2^{M'} + f^{M_1} f_2{}^{M_1'} + \cdots}{f^N}.$$

Since we assume homogeneity in the variables, we have $M = M_1 = \cdots$. Because of homogeneity in the coefficients, we therefore have $M' = M_1' = \cdots$, and finally that $N \leq M$, that is

$$\mathcal{C} = f^\mu f_2^\nu.$$

From this follows:

Theorem *The complete system of forms of a binary quadratic form is given by the base form and the discriminant.*

("System of forms" means the system of invariants and covariants.)

II. $n = 3$. The complete system of invariants and covariants, of which all others are polynomial expressions, is given by

$$f = f_{1,3}, \quad f_2 = \mathcal{H}_{2,2}, \quad f_3 = \mathcal{J}_{3,3}, \quad d = d_{4,0},$$

where d is the discriminant of f (Lecture XVII), even though we cannot yet prove this here. According to our theorem, d must be a rational function of f, f_2, f_3, that is, we must have a syzygy between f, $\mathcal{H}$, $\mathcal{J}$, and d, which we can easily determine from our theorem. In general, denote by $\bar{q}$ the result of setting $a_1 = 0$ in q. If we set $a_1 = 0$ (Lecture XVII) in

$$d = a_0^2a_3^2 - 3a_1^2a_2^2 + 4a_1^3a_3 + 4a_0a_2^3 - 6a_0a_1a_2a_3,$$

then

$$\bar{d} = a_0^2a_3^2 + 4a_0a_2^3,$$

and, furthermore,

$$\bar{c}_2 = a_0 a_2, \qquad \text{whence } a_2 = \frac{\bar{c}_2}{a_0},$$

$$\bar{c}_3 = a_0^2 a_3, \qquad \text{whence } a_3 = \frac{\bar{c}_3}{a_0^2}.$$

From this it follows that

$$\begin{aligned}\bar{d} &= a_0^2 \frac{\bar{c}_3^2}{a_0^4} + 4a_0 \frac{\bar{c}_2^3}{a_0^3} \\ &= \frac{\bar{c}_3^2 + 4\bar{c}_2^3}{a_0^2},\end{aligned}$$

so our theorem implies that

$$d = \frac{f_3^2}{f^2} + 4\frac{f_2}{f^2}$$

and the desired syzygy is

$$4\mathcal{H}^3 = df^2 - \mathcal{J}^2. \tag{A}$$

This relation is very important. We remark that it gives, at the same time, the representation of the square of the unique skew covariant $\mathcal{J}$ as a polynomial function of the others, which subsequently will occur often.

Lecture XXI (June 15, 1897)

The syzygy (A) gives the solution of the cubic equation. This solution, which we shall study in detail subsequently, shows a connection between covariants and the theory of equations. We give here an exact, rigorous, and systematic treatment of the solution of the cubic equation $f = 0$. We distinguish two cases according to whether the equation $f = 0$ does or does not have multiple roots.

1. $d \neq 0$. We have

$$\mathcal{H}^3 = \tfrac{1}{4}(\sqrt{d}f - \mathcal{J})(\sqrt{d}f + \mathcal{J}).$$

In this case $(\sqrt{d}f - \mathcal{J})$ and $(\sqrt{d}f + \mathcal{J})$ cannot have a common factor. Because, if this were the case, then it would also have to occur in f and $\mathcal{J}$. If it is $x_1 - \alpha x_2$, for instance, then we apply the transformation

$$\begin{aligned}x_1 - \alpha x_2 &= x_2', \\ x_1 &= x_1'.\end{aligned}$$

Then, if $f = f' = a_0' x_1'^3 + 3a_1' x_1'^2 x_2' + \cdots$, then $\mathcal{H}$, $\mathcal{J}$, d, f are unchanged by the transformation, up to a power of the substitution determinant, because of their covariant property. Since f', $\mathcal{J}'$, and hence $\mathcal{H}'$, are supposed to contain the factor x_2', we have

$$a_0' = 0,$$
$$a_0' a_2' - a_1'^2 = 0,$$

whence

$$a_1' = 0.$$

Thus, f' would contain the factor twice, and thus so would f, which contradicts the hypothesis. But we want to emphasize once again, how essential the covariant property of f, $\mathcal{H}$, $\mathcal{J}$, d is in this argument, because from it alone did we conclude that the covariants $\mathcal{H}'$, $\mathcal{J}'$, d' belonging to f' are essentially identical to $\mathcal{H}$, $\mathcal{J}$, d, and that therefore a factor common to f and $\mathcal{J}$ also has to be a common factor of f' and $\mathcal{J}'$, after applying an appropriate transformation. However, if $\alpha = 0$, then one cannot apply the transformation; then f, $\mathcal{J}$, $\mathcal{H}$ have the factor x_1 in common, so $a_3 = 0$, $a_1 a_3 - a_2^2 = 0$, and therefore $a_2 = 0$. But then f contains the factor x_1 twice, which again is impossible.

Thus, if $d \neq 0$, then the functions $(\sqrt{d} f - \mathcal{J})$ and $(\sqrt{d} f + \mathcal{J})$ have no common factors. So, if we set

$$\mathcal{H} = l \cdot m,$$

where $l \neq m$ because of the last fact, then necessarily

$$\frac{1}{2}(\sqrt{d} f - \mathcal{J}) = l^3,$$
$$\frac{1}{2}(\sqrt{d} f + \mathcal{J}) = m^3,$$

from which we conclude that

$$f = \frac{1}{\sqrt{d}}(l + m)\,(l + \epsilon m)\,(l + \epsilon^2 m),$$

with $1 \neq \epsilon = \sqrt[3]{1}$. To factor f one therefore has to either first factor $\mathcal{H}$ and form $l + \epsilon^i m$, or, if one does not even want to solve a quadratic equation, one has to form $\mathcal{J}$ and take the cube roots of $\frac{1}{2}(\sqrt{d} f + \mathcal{J})$ and $\frac{1}{2}(\sqrt{d} f - \mathcal{J})$, which has to be possible, and then proceed as before.

2. $d = 0$, $\mathcal{H} \neq 0$. In this case the syzygy reduces to

$$\mathcal{J}^2 = -4\mathcal{H}^3.$$

After applying a suitable transformation, we may assume that f contains the factor x_2 twice; therefore $a_0 = 0$ and $a_1 = 0$. If we apply the transformation to $\mathcal{H}$ also, then we must get the covariant $\mathcal{H}'$ of the transformed form, which contains the factor x_2 twice, since $a_0 = 0$, $a_1 = 0$. Thus, it also was contained in it twice before the transformation. We therefore only need to take the square root of $\mathcal{H}$—which is possible—and obtain the factor which occurs twice in f. The third factor, which can then be found immediately, must be different from these, otherwise—always after an appropriate transformation—we would have $a_0 = 0$, $a_1 = 0$, $a_2 = 0$, and therefore $\mathcal{H} \equiv 0$, which contradicts our assumption.

3. $d = 0$, $\mathcal{H} \equiv 0$ (identically in the x); therefore, $\mathcal{J} \equiv 0$ as well. Suppose f contains the factor x_2 twice, that is—we assume that the transformation has already been applied—$a_0 = 0$, $a_1 = 0$; since $\mathcal{H} \equiv 0$ we have $a_1 a_3 - a_2^2 = 0$, so in particular $a_2 = 0$. Therefore, f is the third power of a linear form, and one only has to take the third root of f.

We summarize the three cases.

1. $d \neq 0$. Three distinct roots.
2. $d = 0$, $\mathcal{H} \neq 0$. Two distinct roots, one a double root.
3. $d = 0$, $\mathcal{H} = 0$. One triple root.

One can see from this calculation how closely connected the vanishing of the invariants and the identical vanishing of the covariants is with the type of solutions of the equation $f = 0$. The key factor in determining the type of solution was the existence of certain covariants of f, that is, functions that are related to f in a way that cannot be destroyed by linear transformations, so that, as long as one only considers these functions, one may assume without loss of generality that any appearing linear factors are equal to x_1 or x_2.

Using these same principles one can solve the equation of degree four. We will return to this and will see that it is also based on a syzygy, just as the cubic equation.

Lecture XXII (June 17, 1897)

III. $n = 4$. The complete system of forms for a form of order four is given by the following five in- and covariants, whose invariant property we have proven earlier (Lectures XI, XII, XIX):

$$\begin{aligned}
f = f_{1,4} &= a_0x_1^4 + 4a_1x_1^3x_2 + 6a_2x_1^2x_2^2 + 4a_3x_1x_2^3 + a_4x_2^4,\\
f_2 = \mathcal{H}_{2,4} &= (a_0a_2 - a_1^2)x_1^4 + \cdots,\\
f_3 = \mathcal{J}_{3,6} &= (a_0^2a_3 - 3a_0a_1a_2 + 2a_1^3)x_1^6 + \cdots,\\
f_4 = i_{2,0} &= a_0a_4 - 4a_1a_3 + 3a_2^2,\\
j_{3,0} &= a_0a_2a_4 + 2a_1a_2a_3 - a_2^3 - a_0a_3^2 - a_1^2a_4.
\end{aligned}$$

Only later will we be able to show that all remaining in- and covariants are polynomial expressions in these. According to our general theorem, however, all in- and covariants can be expressed rationally in terms of f, $\mathcal{H}$, $\mathcal{J}$, i. So, in particular, j can be expressed rationally in terms of these, that is, there is a syzygy between our five covariants. With our general theorem it can be determined without difficulty. We have

$$\bar{j} = a_0a_2a_4 - a_2^3 - a_0a_3^2,$$

and furthermore

$$\bar{c}_4 = a_0a_4 + 3a_2^2;$$

hence

$$a_4 = \frac{\bar{c}_4 - 3a_2^2}{a_0} = \frac{\bar{c}_4 - 3\frac{\bar{c}_2^2}{a_0^2}}{a_0} = \frac{\bar{c}_4a_0^2 - 3\bar{c}_2^2}{a_0^3},$$

$$\bar{c}_3 = a_0^2a_3$$

and

$$a_3 = \frac{\bar{c}_3}{a_0^2},$$

$$\bar{c}_2 = a_0a_2;$$

therefore

$$a_2 = \frac{\bar{c}_2}{a_0}.$$

Thus, it follows that

$$j = f\,\frac{f_2}{f}\,\frac{f_4f^2 - 3f_2^2}{f^3} - \frac{f_2^3}{f^3} - f\,\frac{f_3^2}{f^4},$$

or

$$jf^3 = \mathcal{H}f^2i - 3\mathcal{H}^3 - \mathcal{H}^3 - \mathcal{J}^2,$$

or finally

$$jf^3 - \mathcal{H}f^2i + 4\mathcal{H}^3 + \mathcal{J}^2 = 0. \tag{B}$$

This syzygy is of degree six and order twelve. Here too, one can interpret the syzygy as a polynomial representation of the square of the only skew invariant $\mathcal{J}$ in terms of f, $\mathcal{H}$, i, j; in fact, it is very common that the square of a skew in- or covariant is a polynomial expression of the remaining in- and covariants that together constitute the full system of forms.

The equation in x

$$-\mathcal{J}^2 = 0$$

is identical to the cubic equation for $\frac{\mathcal{H}}{f}$

$$4\mathcal{H}^3 - i\,f^2\mathcal{H} + j f^3 = 0.$$

If we set $\frac{\mathcal{H}}{f} = \lambda$, then it reads

$$4\lambda^3 - i\lambda + j = 0. \tag{1}$$

Let the roots of this equation be λ_1, λ_2, λ_3; then

$$4\lambda^3 - i\lambda + j = 4(\lambda - \lambda_1)(\lambda - \lambda_2)(\lambda - \lambda_3).$$

And therefore we can write:

$$\mathcal{J}^2 = -4(\mathcal{H} - \lambda_1 f)(\mathcal{H} - \lambda_2 f)(\mathcal{H} - \lambda_3 f). \tag{2}$$

This formula now also allows us to solve the fourth degree equation $f = 0$. We only treat the case $d \neq 0$ in some detail, however. It is well known that in the case of multiple roots the analysis reduces to equations of lower degree, and we can then apply the methods discussed for the cubic equation. For $d = 0$, we only note what the exceptional cases are (d is the discriminant of f).

I. $d \neq 0$. First of all, in this case the three factors $\mathcal{H} - \lambda_i f$ must be mutually distinct, that is, $\lambda_1 \neq \lambda_2 \neq \lambda_3$. Furthermore, no two of the quantities $\mathcal{H} - \lambda_i f$ can have a common factor. Otherwise, this factor would also appear in $\mathcal{H}$ and f. If it is x_2, then $a_0 = 0$, and because $a_0a_2 - a_1^2 = 0$, we must also have $a_1 = 0$. But this is not possible because then $d = 0$. Thus, equation (1) must have three distinct roots; hence its discriminant d' must be nonzero. But

$$d' = 27j^2 - i^3,$$

and must be identical to d up to a constant. Hence, if we can prove this, then the claim $\lambda_1 \neq \lambda_2 \neq \lambda_3$ will also be proven. The discriminant of a form of order n is of degree $2(n-1)$ in the coefficients, hence d has degree six, just like d'. Furthermore, d is an irreducible function.

If $d = 0$, then f has a double factor; after an appropriate transformation, we may assume this factor to be x_2. Then $a_0 = 0$, $a_1 = 0$; therefore

$$j(0,0,a_2,a_3,a_4) = -a_2^3,$$
$$i(0,0,a_2,a_3,a_4) = 3a_2^2,$$

and thus it follows that

$$d'(0,0,a_2,a_3,a_4) = 27a_2^6 - 27a_2^6 = 0,$$

which, therefore, also has to be the case before any transformation is applied. Whenever $d = 0$, then also $d' = 0$; this means that d is a factor of d', and since both have the same degree they can only differ by a constant. Thus we have found at the same time that the discriminant of a biquadratic form is

$$d = 27j^2 - i^3.$$

From this we conclude that $\mathcal{H} - \lambda_i f$ has to be a square, that is,

$$\mathcal{H} - \lambda_1 f = \phi_1^2, \qquad \mathcal{H} - \lambda_2 f = \phi_2^2, \qquad \mathcal{H} - \lambda_3 f = \phi_3^2,$$

where no two of the functions ϕ_1, ϕ_2, ϕ_3 have a common factor. Finally, it therefore follows that

$$(\lambda_2 - \lambda_1)f = (\phi_1 + \phi_2)(\phi_1 - \phi_2),$$
$$(\lambda_3 - \lambda_2)f = (\phi_2 + \phi_3)(\phi_2 - \phi_3),$$
$$(\lambda_1 - \lambda_3)f = (\phi_3 + \phi_1)(\phi_3 - \phi_1).$$

Lecture XXIII (June 18, 1897)

These three decompositions into quadratic factors are the only three possible ones, that is, they are all distinct, and one has

$$\phi_1 + \phi_2 = l_1 l_2 c_3; \qquad \phi_2 + \phi_3 = l_1 l_3 c_1; \qquad \phi_3 + \phi_1 = l_2 l_3 c_2;$$
$$\phi_1 - \phi_2 = l_3 l_4 c_3'; \qquad \phi_2 - \phi_3 = l_2 l_4 c_1'; \qquad \phi_3 - \phi_1 = l_1 l_4 c_2'.$$

where the l_i are the linear factors of f and the c_j, c_j' are constants. To prove that the decompositions are indeed different from one another, we observe that

$$\begin{vmatrix} 1 & \lambda_1 & \phi_1^2 \\ 1 & \lambda_2 & \phi_2^2 \\ 1 & \lambda_3 & \phi_3^2 \end{vmatrix} = 0$$

holds identically because of the defining equations for ϕ_1, ϕ_2, ϕ_3; thus, we also have

$$(\lambda_3 - \lambda_2)\phi_1^2 + (\lambda_1 - \lambda_3)\phi_2^2 + (\lambda_2 - \lambda_1)\phi_3^2 = 0.$$

If there existed a linear relation between ϕ_1, ϕ_2, ϕ_3, then we could set $\phi_3 = \alpha_1\phi_1 + \alpha_2\phi_2$; there would therefore exist a quadratic relation for ϕ_1, ϕ_2:

$$\beta_1(\phi_1 - \mu_1\phi_2)(\phi_1 - \mu_2\phi_2) = 0,$$

holding identically in x. Then ϕ_1 and ϕ_2 would only differ by a constant factor, which cannot be the case because then $\lambda_1 = \lambda_2$. Now, if $\phi_1 + \phi_2$ equaled $\phi_2 + \phi_3$ or $\phi_2 - \phi_3$ up to a constant factor, then there would exist a linear relation between ϕ_1, ϕ_2, ϕ_3. Since one can make an analogous argument for the other factors, we have shown that the three decompositions are mutually distinct.

It is apparent from the above that the solution of the fourth degree equation $f = 0$ rests in an essential way on the solution of the cubic equation $4\lambda^3 - i\lambda + j = 0$, the so-called *resolvent cubic.* After solving it one forms $\mathcal{H} - \lambda_i f$, takes square roots—which must be possible—and obtains in this way ϕ_1, ϕ_2, ϕ_3 whose pairwise sums and differences are the factors of f. One does not even have to solve the quadratic equations $\phi_1+\phi_2 = 0$, etc.; it is sufficient to determine the greatest common divisor of $\phi_1+\phi_2$ and $\phi_2+\phi_3$, etc. From this one obtains one factor from which one can easily determine the others.

At the same time, we have accomplished the factorization of $\mathcal{H}$; we have

$$(\lambda_2 - \lambda_1)\mathcal{H} = \lambda_2\left(\phi_1 + \sqrt{\frac{\lambda_1}{\lambda_2}}\phi_2\right)\left(\phi_1 - \sqrt{\frac{\lambda_1}{\lambda_2}}\phi_2\right),$$

$$(\lambda_3 - \lambda_2)\mathcal{H} = \lambda_3\left(\phi_2 + \sqrt{\frac{\lambda_2}{\lambda_3}}\phi_3\right)\left(\phi_2 - \sqrt{\frac{\lambda_2}{\lambda_3}}\phi_3\right),$$

$$(\lambda_1 - \lambda_3)\mathcal{H} = \lambda_1\left(\phi_3 + \sqrt{\frac{\lambda_3}{\lambda_1}}\phi_1\right)\left(\phi_3 - \sqrt{\frac{\lambda_3}{\lambda_1}}\phi_1\right),$$

and here too we obtain all three possible factorizations—for the same reasons as before.

Since

$$\mathcal{J} = \sqrt{-4}\,\phi_1\phi_2\phi_3,$$

one realizes that the equation $\mathcal{J} = 0$ belongs to those equations of degree six that are solvable by radicals. Thus, it obviously belongs to a special class of equations of degree six.

Finally, we remark that the quantities λ_1, λ_2, λ_3 belong to the so-called irrational invariants; likewise, the ϕ_1, ϕ_2, ϕ_3 belong to the so-called irrational covariants, expressions that are self-explanatory. The invariants λ_i have weight two, as is easily seen; the covariants ϕ_i have weight one.

We briefly consider the example:

$$f = x_1^4 + x_2^4;$$

therefore, $a_0 = a_4 = 1$, $a_1 = a_2 = a_3 = 0$. Here, we have

$$\begin{aligned}\mathcal{H} &= x_1^2 x_2^2,\\ \mathcal{J} &= x_1 x_2 (x_1^4 - x_2^4),\\ i &= 1,\\ j &= 0,\\ d &= -1.\end{aligned}$$

The resolvent cubic is

$$4\lambda^3 - \lambda = 0.$$

Therefore

$$\lambda_1 = 0, \qquad \lambda_2 = \frac{1}{2}, \qquad \lambda_3 = -\frac{1}{2};$$

$$\begin{aligned}\mathcal{H} - \lambda_1 f &= (x_1 x_2)^2, & \phi_1 &= x_1 x_2;\\ \mathcal{H} - \lambda_2 f &= -\frac{1}{2}(x_1^2 - x_2^2)^2, & \phi_2 &= (x_1^2 - x_2^2)\sqrt{-\frac{1}{2}};\\ \mathcal{H} - \lambda_3 f &= \frac{1}{2}(x_1^2 + x_2^2)^2, & \phi_3 &= (x_1^2 + x_2^2)\sqrt{\frac{1}{2}}.\end{aligned}$$

This can be used to readily verify the theorems derived above.

We only list the results in the case where f has multiple linear factors.

II. $d = 0$.

1. $i \neq 0$; thus $j \neq 0$, $\mathcal{J} \neq 0$. Then we have

$$\lambda_1 = \lambda_2 \neq \lambda_3,$$
$$f = l_1^2 l_3 l_4, \qquad l_1 \neq l_3 \neq l_4,$$
$$\phi_1 = \phi_2 = c l_1^2,$$

ϕ_3 contains the factor l_1 once,

$$\mathcal{J} = l_1^5 m.$$

The double root can be found by taking the square root of ϕ_1.

2. $i = 0$; thus $j = 0$, $\mathcal{J} \neq 0$, $\mathcal{H} \neq 0$, and

$$\mathcal{J}^2 = -4\mathcal{H}^3.$$

$\mathcal{J}$ is the third power and $\mathcal{H}$ is the square of an expression of order two. Furthermore,

$$\lambda_1 = \lambda_2 = \lambda_3 = 0,$$
$$f = l_1^3 l_3,$$
$$\phi_1 = \phi_2 = \phi_3 = l_1^2,$$
$$\mathcal{H} = l_1^4,$$
$$\mathcal{J} = l_1^6.$$

3. $i \neq 0$; $\mathcal{J} \equiv 0$.

$$f = c\mathcal{H} = (l_3 l_4)^2,$$
$$\mathcal{H} = \phi_3^2,$$
$$\phi_3 = l_3 l_4,$$
$$\phi_1 = 0,$$
$$\phi_2 = 0.$$

4. $i = 0$; $j = 0$; $\mathcal{H} \equiv 0$; $\mathcal{J} \equiv 0$.

$$\phi_1 = \phi_2 = \phi_3 = 0,$$
$$f = l_1^4.$$

The four cases can be characterized as follows: In case (1) there is one invariant, whereas there are none in case (2); but there are two covariants in both cases; in case (3), there is only one invariant and one covariant; and in case (4) there exist neither invariants nor covariants.

As has been done for forms of order three and four, one could investigate for forms of order five and higher how the in- and covariants are related to the properties of the equation $f = 0$, with respect to multiple roots, etc. While there are some results in this direction, there is not yet a complete theory, even for forms of order five. From here on, of course, the difference is that such a decomposition into linear factors, which could be done for forms of order three and four, is not possible any more in general; but even there the solution of the equations was only an incidental result, as we pointed out from the beginning.

Lecture XXIV (June 21, 1897)

I.9 Simultaneous invariants and covariants

So far we have always begun our observations with one base form. But most of our results can be greatly generalized if we begin with an arbitrary number of forms, that is, if we consider a system of "simultaneous forms." If we are given a number of forms, say two (the extension to arbitrarily many forms will be self-evident) forms of orders n and m

$$f^{(n)}(x_1, x_2) = a_0 x_1^n + \binom{n}{1} a_1 x_1^{n-1} x_2 + \cdots + a_n x_2^n,$$

$$g^{(m)}(x_1, x_2) = b_0 x_1^m + \binom{m}{1} b_1 x_1^{m-1} x_2 + \cdots + b_m x_2^m,$$

then we want to apply *the same* linear transformation to both of them:

$$x_1 = \alpha_{11} x_1' + \alpha_{12} x_2',$$
$$x_2 = \alpha_{21} x_1' + \alpha_{22} x_2'.$$

Then f, respectively g, gets transformed into a new form f' of order n, respectively g' of order m. Let

$$f'^{(n)}(x_1', x_2') = a_0' {x_1'}^n + \binom{n}{1} a_1' {x_1'}^{n-1} x_2' + \cdots + a_n' {x_2'}^n,$$

$$g'^{(m)}(x_1', x_2') = b_0' {x_1'}^m + \binom{m}{1} b_1' {x_1'}^{m-1} x_2' + \cdots + b_m' {x_2'}^m.$$

Analogously to the above, we then define a *simultaneous invariant* of f and g to be a polynomial function $\mathcal{I}$ of the coefficients a and b that satisfies the equation

$$\mathcal{I}(a_0', a_1', \ldots, a_n'; \; b_0', b_1', \ldots, b_m') = \delta^p \mathcal{I}(a_0, a_1, \ldots, a_n; \; b_0, b_1, \ldots, b_m).$$

The exponent p is called the weight of the invariant. Similarly, we define a *simultaneous covariant* of f and g to be a polynomial function $\mathcal{C}$ of the coefficients a and b and the variables x that satisfies the equation

$$\begin{aligned}\mathcal{C}(a_0', a_1', \ldots, a_n';\ b_0', b_1', \ldots, b_m';\ x_1', x_2') \\ = \delta^p \mathcal{C}(a_0, a_1, \ldots, a_n;\ b_0, b_1, \ldots, b_m;\ x_1, x_2).\end{aligned}$$

The exponent p is called the weight of the covariant. The usual in- and covariants are a special case of these; they are obtained when the "degree" in the b (or the a) becomes zero. The generalization to more base forms is evident. For instance, if $n = 2$, $m = 2$, then the function

$$\mathcal{I} = a_0 b_2 - 2a_1 b_1 + a_2 b_0$$

is an invariant. This is easily seen if one substitutes into $a_0' b_2' + 2a_1' b_1' + a_2' b_0'$ the expressions (Lecture VI):

$$\begin{aligned}a_0' &= a_0 \alpha_{11}^2 + 2a_1 \alpha_{11} \alpha_{21} + a_2 \alpha_{21}^2, \\ a_1' &= a_0 \alpha_{11} \alpha_{12} + a_1(\alpha_{11}\alpha_{22} + \alpha_{12}\alpha_{21}) + a_2 \alpha_{21} \alpha_{22}, \\ a_2' &= a_0 \alpha_{12}^2 + 2a_1 \alpha_{12} \alpha_{22} + a_2 \alpha_{22}^2;\end{aligned}$$

$$\begin{aligned}b_0' &= b_0 \alpha_{11}^2 + 2b_1 \alpha_{11} \alpha_{21} + b_2 \alpha_{21}^2, \\ b_1' &= b_0 \alpha_{11} \alpha_{12} + b_1(\alpha_{11}\alpha_{22} + \alpha_{12}\alpha_{21}) + b_2 \alpha_{21} \alpha_{22}, \\ b_2' &= b_0 \alpha_{12}^2 + 2b_1 \alpha_{12} \alpha_{22} + b_2 \alpha_{22}^2.\end{aligned}$$

We see just as before, when we showed (Section I.4, Lecture VII) that we may assume invariants and covariants to be homogeneous, that if $\mathcal{C}$ has the invariant property, then all the parts into which $\mathcal{C}$ decomposes also have the invariant property, provided that each part individually is homogeneous in the x, as well as the a, the b, etc. Without loss of generality one may, therefore, assume—which we want to do—that all expressions to be considered are homogeneous in all coefficient sequences, as well as the variables.

We can now immediately extend all the properties of invariants and covariants that we derived earlier (Section I.4 ff.). We indicate this briefly for the case of two base forms. Let

$$\mathcal{I} = \sum Z a_0^{\nu_0} a_1^{\nu_1} \cdots a_n^{\nu_n} b_0^{\mu_0} b_1^{\mu_1} \cdots b_m^{\mu_m}$$

be a simultaneous invariant of f and g. After applying transformation (1) (Lectures V, VII):

$$\begin{aligned} x_1 &= \kappa x_1', \\ & \qquad\qquad (\delta = \kappa\lambda), \\ x_2 &= \lambda x_2', \end{aligned} \tag{1}$$

the transformed coefficients are

$$\begin{aligned} a_0' &= a_0\kappa^n, & b_0' &= b_0\kappa^m, \\ &\dots & &\dots \\ a_i' &= a_i\kappa^{n-i}\lambda^i, & b_i' &= b_i\kappa^{m-i}\lambda^i, \\ &\dots & &\dots \\ a_n' &= a_n\lambda^n, & b_m' &= b_m\lambda^m, \end{aligned}$$

and we must therefore have that

$$\begin{aligned} &\sum Za_0^{\nu_0}a_1^{\nu_1}\cdots a_n^{\nu_n}\kappa^{n\nu_0+(n-1)\nu_1+\cdots+\nu_{n-1}+m\mu_0+(m-1)\mu_1+\cdots+\mu_{m-1}} \\ &\qquad \times b_0^{\mu_0}b_1^{\mu_1}\cdots b_m^{\mu_m}\lambda^{\nu_1+2\nu_2+\cdots+n\nu_n+\mu_1+2\mu_2+\cdots+m\mu_m} \\ &\qquad = \kappa^p\lambda^p\sum Za_0^{\nu_0}a_1^{\nu_1}\cdots a_n^{\nu_n}b_0^{\mu_0}b_1^{\mu_1}\cdots b_m^{\mu_m}, \end{aligned}$$

where p is the weight of the invariant. Hence, it follows that for each individual term we must have:

$$\begin{gathered} n\nu_0 + (n-1)\nu_1 + \cdots + \nu_{n-1} + m\mu_0 + (m-1)\mu_1 + \cdots + \mu_{m-1} = p, \\ \nu_1 + 2\nu_2 + \cdots + n\nu_n + \mu_1 + 2\mu_2 + \cdots + m\mu_m = p. \end{gathered}$$

Addition of these two formulas gives

$$n(\nu_0 + \nu_1 + \cdots + \nu_n) + m(\mu_0 + \mu_1 + \cdots + \mu_m) = 2p,$$

or, if

$$\begin{gathered} \nu_0 + \nu_1 + \cdots + \nu_n = r, \\ \mu_0 + \mu_1 + \cdots + \mu_m = s \end{gathered}$$

are the degrees in the a, respectively the b, then

$$nr + ms = 2p.$$

This formula and the following

$$\nu_1 + 2\nu_2 + \cdots + n\nu_n + \mu_1 + 2\mu_2 + \cdots + m\mu_m = p$$

(which, incidentally, provides a new definition of the weight) are the necessary and sufficient conditions for the function $\mathcal{I}$ to be an invariant with respect to the transformation (1). The weight of every term must be constant, namely, equal to $\frac{1}{2}(nr+ms)$.

The situation is analogous for the covariants. If

$$\mathcal{C} = C_0 x_1^M + \binom{M}{1} C_1 x_1^{M-1} x_2 + \cdots$$

is such a covariant, and if p is the weight of C_0, then we find for the weight of a term C_i, which has to be the same for all its summands:

$$p_i = p + i,$$

and furthermore

$$nr + ms - 2p = M.$$

The weight of C_0 is at the same time the weight of the covariant. The presence of these properties is at the same time sufficient for invariance with respect to (1).

Everything remains valid for arbitrarily many base forms; one only needs to set

$$\begin{aligned} p &= \nu_1 + 2\nu_2 + \cdots + n\nu_n + \mu_1 + 2\mu_2 + \cdots + m\mu_m + \cdots , \\ 2p &= nr + ms + \cdots , \end{aligned}$$

and

$$M = nr + ms + \cdots - 2p.$$

The properties that result from invariance with respect to (2):

$$\begin{aligned} x_1 &= x_1' + \mu x_2', \\ x_2 &= \qquad\ x_2' \end{aligned} \qquad (\delta = 1), \qquad (2)$$

can be derived equally easily. The transformed coefficients are (Lecture VIII):

$$\begin{aligned} a_i' &= a_i + \binom{i}{1}\mu a_{i-1} + \binom{i}{2}\mu^2 a_{i-2} + \cdots + \mu^i a_0, \\ b_i' &= b_i + \binom{i}{1}\mu b_{i-1} + \binom{i}{2}\mu^2 b_{i-2} + \cdots + \mu^i b_0, \end{aligned}$$

where

$$\frac{da_i'}{d\mu} = i a_{i-1}', \qquad \frac{db_i'}{d\mu} = i b_{i-1}'.$$

Since $\delta = 1$, an invariant $\mathcal{I}$ has to identically satisfy the equation

$$\mathcal{I}(a_0', a_1', \ldots, a_n';\ b_0',\ b_1', \ldots, b_m') = \mathcal{I}(a_0, a_1, \ldots, a_n;\ b_0,\ b_1, \ldots, b_m).$$

Differentiation of this equation with respect to μ gives:

$$\begin{aligned} &\frac{\partial \mathcal{I}(a';b')}{\partial a_1'} a_0' + \frac{\partial \mathcal{I}(a';b')}{\partial a_2'} 2a_1' + \cdots + \frac{\partial \mathcal{I}(a';b')}{\partial a_n'} n a_{n-1}' \\ &+ \frac{\partial \mathcal{I}(a';b')}{\partial b_1'} b_0' + \frac{\partial \mathcal{I}(a';b')}{\partial b_2'} 2b_1' + \cdots + \frac{\partial \mathcal{I}(a';b')}{\partial b_m'} m\, b_{m-1}' = 0. \end{aligned}$$

If we now set

$$\begin{aligned} \mathbf{D}_a &= a_0 \frac{\partial}{\partial a_1} + 2a_1 \frac{\partial}{\partial a_2} + \cdots + n a_{n-1} \frac{\partial}{\partial a_n}, \\ \mathbf{D}_b &= b_0 \frac{\partial}{\partial b_1} + 2b_1 \frac{\partial}{\partial b_2} + \cdots + m b_{m-1} \frac{\partial}{\partial b_m}, \\ \mathbf{D} &= \mathbf{D}_a + \mathbf{D}_b, \end{aligned}$$

then an invariant $\mathcal{I}$ satisfies the differential equation

$$\mathbf{D}\mathcal{I} = 0.$$

The converse can also be shown analogously to before (Lecture VIII). Finally, it is obvious that an invariant $\mathcal{I}$ also has to satisfy the differential equation

$$\mathbf{\Delta}\mathcal{I} = 0,$$

where

$$\mathbf{\Delta} = \mathbf{\Delta}_a + \mathbf{\Delta}_b,$$

and

$$\begin{aligned} \mathbf{\Delta}_a &= n a_1 \frac{\partial}{\partial a_0} + (n-1) a_2 \frac{\partial}{\partial a_1} + \cdots + a_n \frac{\partial}{\partial a_{n-1}}, \\ \mathbf{\Delta}_b &= m b_1 \frac{\partial}{\partial b_0} + (m-1) b_2 \frac{\partial}{\partial b_1} + \cdots + b_m \frac{\partial}{\partial b_{m-1}}. \end{aligned}$$

This is immediately apparent if one observes that $\mathcal{I}$ also has to be an invariant with respect to type (3):

$$\begin{aligned} x_1 &= x_1', \\ x_2 &= \nu x_1' + x_2'. \end{aligned} \qquad (\delta = 1), \qquad (3)$$

The converse also holds.

Regarding the covariants, they of course have to satisfy the differential equations

$$\mathbf{D}\mathcal{C} = x_2 \frac{\partial \mathcal{C}}{\partial x_1},$$

$$\mathbf{\Delta}\mathcal{C} = x_1 \frac{\partial \mathcal{C}}{\partial x_2},$$

and if the first differential equation is satisfied, then $\mathcal{C}$ is a covariant with respect to (2); if the second one is satisfied, $\mathcal{C}$ is a covariant with respect to (3).

Since a general linear transformation is built from (1), (2), (3), it is clear that the properties we derived for invariants and covariants constitute the necessary and sufficient conditions for $\mathcal{I}$ or $\mathcal{C}$ to be an in- or covariant.

Lecture XXV (June 22, 1897)

But this system of conditions can be reduced in size, just as in the case of one base form. One only needs to prove the pertinent relations between $\mathbf{D}$ and $\mathbf{\Delta}$, which is easily done.

It is self-evident that $\mathbf{D}$ and $\mathbf{\Delta}$ satisfy the usual rules of differentiation with respect to sums, products, etc. (Lecture X). We always want to assume that the expressions to which we apply them are homogeneous of degree r in the coefficients of the form f, of degree s in the coefficients of the form g, etc. Furthermore, they should be isobaric of weight p in both groups of coefficients jointly (we do not assume isobarity in the individual coefficients).

Theorem *The operations $\mathbf{D}$ and $\mathbf{\Delta}$, applied to such an expression, do not change the degree of each coefficient sequence; $\mathbf{D}$ lowers the total weight by 1, $\mathbf{\Delta}$ raises it by 1.*

Most important is the following:

Theorem *If $\mathcal{A}$ is a function of the $a_i, b_i, \ldots$, homogenous of degree r in the a_i, homogeneous of degree s in the $b_i, \ldots$, and isobaric of weight p in coefficient groups jointly, then*

$$(\mathbf{D}\mathbf{\Delta} - \mathbf{\Delta}\mathbf{D})\mathcal{A} = (nr + ms + \cdots - 2p)\mathcal{A}.$$

Because of its importance, we want to prove this theorem. We decompose $\mathcal{A}$ into

$$\mathcal{A} = \mathcal{A}_1 + \mathcal{A}_2 + \cdots ,$$

so that $\mathcal{A}_1$, $\mathcal{A}_2, \ldots$ each are isobaric in the a_i, $b_i, \ldots$; suppose for instance that $\mathcal{A}_1$ is isobaric of weight p_1 in the a_i, of weight p_2 in the b_i, etc. Now, clearly

$$(\mathbf{D}_a\mathbf{\Delta}_b - \mathbf{\Delta}_b\mathbf{D}_a)\mathcal{A}_1 = 0,$$

where a, b are two arbitrary distinct coefficient sequences, since it is irrelevant whether we first differentiate with respect to the a and then the b, or vice versa. Therefore, one has

$$\begin{aligned}(\mathbf{D\Delta} - \mathbf{\Delta D})\mathcal{A}_1 = \{&(\mathbf{D}_a + \mathbf{D}_b + \cdots)(\mathbf{\Delta}_a + \mathbf{\Delta}_b + \cdots) \\ &- (\mathbf{\Delta}_a + \mathbf{\Delta}_b + \cdots)(\mathbf{D}_a + \mathbf{D}_b + \cdots)\}\mathcal{A}_1,\end{aligned}$$

where one is supposed to take all possible combinations, that is, calculate as if multiplying. Therefore,

$$\begin{aligned}(\mathbf{D\Delta} - \mathbf{\Delta D})\mathcal{A}_1 = &(\mathbf{D}_a\mathbf{\Delta}_a - \mathbf{\Delta}_a\mathbf{D}_a)\mathcal{A}_1 + (\mathbf{D}_b\mathbf{\Delta}_b - \mathbf{\Delta}_b\mathbf{D}_b)\mathcal{A}_1 + \cdots \\ &+ \sum_{i\neq k}(\mathbf{D}_i\mathbf{\Delta}_k - \mathbf{\Delta}_k\mathbf{D}_i)\mathcal{A}_1\end{aligned}$$

or

$$(\mathbf{D\Delta} - \mathbf{\Delta D})\mathcal{A}_1 = (\mathbf{D}_a\mathbf{\Delta}_a - \mathbf{\Delta}_a\mathbf{D}_a)\mathcal{A}_1 + (\mathbf{D}_b\mathbf{\Delta}_b - \mathbf{\Delta}_b\mathbf{D}_b)\mathcal{A}_1 + \cdots,$$

or, since the previous theorem holds (Lecture X),

$$\begin{aligned}(\mathbf{D\Delta} - \mathbf{\Delta D})\mathcal{A}_1 &= (nr - 2p_1)\mathcal{A}_1 + (ms - 2p_2)\mathcal{A}_1 + \cdots \\ &= (nr + ms + \cdots - 2p)\mathcal{A}_1.\end{aligned}$$

Since an analogous formula is valid for all $\mathcal{A}_i$, one obtains after addition of all those:

$$(\mathbf{D\Delta} - \mathbf{\Delta D})\mathcal{A} = (nr + ms + \cdots - 2p)\mathcal{A}.$$

From this formula we can derive two others:

$$\begin{aligned}\mathbf{D}^k\mathbf{\Delta} - \mathbf{\Delta D}^k &= k(nr + ms + \cdots - 2p + k - 1)\mathbf{D}^{k-1}, \\ \mathbf{D\Delta}^k - \mathbf{\Delta}^k\mathbf{D} &= k(nr + ms + \cdots - 2p - k + 1)\mathbf{\Delta}^{k-1}.\end{aligned}$$

But since the theorem about when $\mathcal{C}$ is a covariant of a base form makes use only of formulas of this kind, one sees that we have the following very general theorem whose proof is immediate from the observations in Lecture XII.

Theorem *If C_0 is a homogeneous polynomial function in the coefficients a_i of the base form f of degree n, the coefficients b_i of the base form g of degree m, etc., and that is, furthermore, isobaric in all coefficient sequences, of total weight p, and that satisfies the differential equation*

$$\mathbf{D}C_0 = 0,$$

then there is one and only one covariant that has C_0 as a source, which is

$$\mathcal{C} = C_0 x_1^M + \frac{1}{1!}\boldsymbol{\Delta} C_0 x_1^{M-1} x_2 + \frac{1}{2!}\boldsymbol{\Delta}^2 C_0 x_1^{M-2} x_2^2 + \cdots + \frac{1}{M!}\boldsymbol{\Delta}^M C_0 x_2^M.$$

This simultaneous covariant of the forms $f, g, \ldots$ is of degree r in the a_i, of degree s in the $b_i, \ldots$, is homogeneous, and has weight p and order $M = nr + ms + \cdots - 2p$. These conditions are necessary and sufficient. For $M = 0$, we obtain the necessary and sufficient conditions for a simultaneous invariant.

We explain this in terms of an example. Let

$$f^{(3)} = a_0 x_1^3 + 3a_1 x_1^2 x_2 + 3a_2 x_1 x_2^2 + a_3 x_2^3,$$
$$g^{(3)} = b_0 x_1^3 + 3b_1 x_1^2 x_2 + 3b_2 x_1 x_2^2 + b_3 x_2^3.$$

Let $r = 1$, $s = 1$, $M = 0$, and hence $p = 3$. We set

$$\mathcal{I} = \sum Z a_i b_k = Z a_0 b_3 + Z' a_1 b_2 + Z'' a_2 b_1 + Z''' a_3 b_0.$$

Then

$$\mathbf{D}\mathcal{I} = Z' a_0 b_2 + 2Z'' a_1 b_1 + 3a_2 b_1 Z''' + Z'' a_2 b_0 + 2a_1 b_1 Z' + 3b_2 a_0 Z = 0$$

must hold identically, and so it follows that

$$\begin{aligned} Z &= 1 \qquad \text{(by choice)}, \\ Z' &= -3, \\ Z'' &= 3, \\ Z''' &= -1. \end{aligned}$$

Hence there is one and only one simultaneous bilinear invariant of two cubic forms, which is

$$\mathcal{I} = a_0 b_3 - 3a_1 b_2 + 3a_2 b_1 - a_3 b_0.$$

The covariants are really only a special case of simultaneous invariants. We formulate this in the following theorem.

Theorem *A covariant of arbitrarily many forms that is homogeneous in the variables and is of weight p and order M is transformed into a simultaneous invariant of weight $p+M$ of those forms together with the linear form $b_0x_1 + b_1x_2$, if one replaces in it x_1, respectively x_2, by $-b_1$, respectively b_0, and conversely.*

Namely, suppose the transformation formulas are

$$x_1 = \alpha_{11}x_1' + \alpha_{12}x_2',$$
$$x_2 = \alpha_{21}x_1' + \alpha_{22}x_2';$$

hence, if we invert,

$$\delta x_1' = \alpha_{22}x_1 - \alpha_{12}x_2,$$
$$\delta x_2' = -\alpha_{21}x_1 + \alpha_{11}x_2.$$

The linear form above is then transformed into

$$\begin{aligned} f = f' &= b_0(\alpha_{11}x_1' + \alpha_{12}x_2') + b_1(\alpha_{21}x_1' + \alpha_{22}x_2') \\ &= (b_0\alpha_{11} + b_1\alpha_{21})x_1' + (b_0\alpha_{12} + b_1\alpha_{22})x_2', \end{aligned}$$

so

$$b_0' = b_0\alpha_{11} + b_1\alpha_{21},$$
$$b_1' = b_0\alpha_{12} + b_1\alpha_{22};$$

if we solve, then

$$-b_1 = \frac{1}{\delta}(-\alpha_{11}b_1' + \alpha_{12}b_0'),$$

$$b_0 = \frac{1}{\delta}(-\alpha_{21}b_1' + \alpha_{22}b_0').$$

From this one sees that, up to the factor $\frac{1}{\delta}$, x_1 and x_2 can be expressed in terms of x_1' and x_2' in the same way as $-b_1$ and b_0 can be expressed in terms of $-b_1'$, b_0'. Hence, if the covariant is C, then

$$C(\ldots;\, x_1',\, x_2') = \delta^p C(\ldots;\, x_1,\, x_2).$$

If one now substitutes $-b_1$ for x_1, respectively b_0 for x_2, on the right, then one has to substitute $-\frac{b_1'}{\delta}$ for x_1', respectively $\frac{b_0'}{\delta}$ for x_2', on the left. Hence, one obtains

$$C\left(\ldots;\, -\frac{b_1'}{\delta},\, \frac{b_0'}{\delta}\right) = \delta^p C(\ldots;\, -b_1,\, b_0),$$

and, since we assume homogeneity of C,

$$\frac{1}{\delta^M} C(\ldots;\, -b_1',\, b_0') = \delta^p C(\ldots;\, -b_1,\, b_0),$$

which completes the proof of the theorem. Thus, if one does not specify the number of base forms, one is justified in considering only invariants.

This, incidentally, also agrees very well with the differential equations because, since C is a covariant, we have $\mathbf{D}_a C = x_2 \frac{\partial C}{\partial x_1}$, and hence, after the substitution, $\mathbf{D}_a C + b_0 \frac{\partial C}{\partial b_1} = 0$, that is, $\mathbf{D}C = 0$. The situation for the other differential equation is analogous.

Lecture XXVI (June 24, 1897)

We shall omit the extension of the Cayley enumeration method—which apparently has not been done yet—and proceed right away to determine a system of covariants through which all others can be expressed rationally. Aside from the covariants determined in Section I.8, they are the simplest imaginable, namely, those that are linear in the coefficients of two base forms.

Let f be a base form of order n, g one of order m, and $n \geq m$. Then there is one and only one simultaneous covariant of f and g that is homogeneous and linear in the coefficients of both base forms, and that has a given weight $p \leq m$. It has the form:

$$C = \left\{ a_0 b_p - \binom{p}{1} a_1 b_{p-1} + \binom{p}{2} a_2 b_{p-2} - \cdots \right.$$
$$\left. + (-1)^p a_p b_0 \right\} x_1^{n+m-2p} + \cdots, \qquad p \leq m.$$

Using the theorem from the previous lecture, it is easily seen that this is a covariant. The differential equation $\mathbf{D}C_0 = 0$ can be verified by writing the source of C as

$$C_0 = \sum_{i=0}^{p} (-1)^i a_i b_{p-i} \binom{p}{i}$$

and applying the operation $\mathbf{D}$ inside the summation sign. This covariant is very important. It is called the *pth transvection of f over g* and is abbreviated as

$$(f, g)_p.$$

The pth transvection of g over f can differ from that of f over g at most by a sign.

The transvection process is a fundamental process for the construction of covariants because one can show—which is not possible here—that if one has a system of base forms and if one forms all possible transvections and transvections of transvections and base forms, and continues in this way, then one obtains all existing covariants.*

The covariants of degree two, which we determined earlier, are indeed the transvections of the form f over itself. For odd p, they vanish identically, and for even p, one has, as one can see easily,

$$f_p = \tfrac{1}{2}(f, f)_p.$$

The covariants $f_3, f_5, \ldots$ are also such transvections, namely, the transvections of the covariants $f_2, f_4, \ldots$ over the base form.

The first transvection of two forms

$$(f, g)_1 = (a_0 b_1 - a_1 b_0) x_1^{n+m-2} + \cdots$$

is especially important. It is also called the *functional determinant* or *Jacobian covariant* of the two forms; one can—up to a constant factor—write it in the form

$$\begin{vmatrix} \dfrac{\partial f}{\partial x_1} & \dfrac{\partial f}{\partial x_2} \\ \dfrac{\partial g}{\partial x_1} & \dfrac{\partial g}{\partial x_2} \end{vmatrix}.$$

The second transvection of a form over itself

$$\tfrac{1}{2}(f, f)_2 = (a_0 a_2 - a_1^2) x_1^{2n-4} + \cdots$$

is also called the *Hessian covariant* of the form; it is—up to a constant factor—the determinant of the second-order derivatives:

$$\begin{vmatrix} \dfrac{\partial^2 f}{\partial x_1^2} & \dfrac{\partial^2 f}{\partial x_1 \partial x_2} \\ \dfrac{\partial^2 f}{\partial x_2 \partial x_1} & \dfrac{\partial^2 f}{\partial x_2^2} \end{vmatrix}.$$

Furthermore, we can now describe how a simultaneous covariant can be expressed as a rational function of transvections and the base forms, in the following theorem.

Theorem *Every simultaneous covariant of the base forms f and g of orders n and m, respectively, where $m \leq n$, can be expressed, up to a power of f that occurs in the denominator, as a polynomial function*

* See Gordan (1868).

of the transvections $f,\ f_2,\ f_3,\ldots,f_n,\ g,\ s_1,\ s_2,\ldots,s_m$, *where in general we set*

$$s_i = (f,\ g)_i.$$

This is done in a manner analogous to that in the theorems (Lecture XX) concerning one base form, if we add

$$b_0 = g,$$

$$b_p = \frac{1}{f}\left\{s_p - \binom{p}{2}a_2b_{p-2} + \binom{p}{3}a_3b_{p-3} - \cdots\right\}$$

to the formulas for a_p *above.*

The proof of the theorem is entirely analogous to that for one base form. Starting with b_m, one has to replace the b_i by the sources of the s_i and, subsequently, the a_i by the sources of the f_i. Now, the argument proceeds as before.

We apply the theorem to simultaneous forms of low order.

1. $n = 1$, $m = 1$. All invariant expressions are polynomial functions of the in- and covariants:

$$f = a_0x_1 + a_1x_2, \qquad g = b_0x_1 + b_1x_2, \qquad s_1 = a_0b_1 - a_1b_0.$$

This is because, according to our theorem, any such C can be expressed in the form:

$$C = \frac{c_0 f^N g^M s_1^{\mathcal{P}} + c_1 f^{N_1} g^{M_1} s_1^{\mathcal{P}_1} + \cdots}{f^Q}.$$

But, since we may assume homogeneity everywhere, we have

$$\begin{aligned} N + M &= N_1 + M_1 = \cdots, \\ M + \mathcal{P} &= M_1 + \mathcal{P}_1 = \cdots, \\ N + \mathcal{P} &= N_1 + \mathcal{P}_1 = \cdots, \end{aligned}$$

from which it follows that

$$N = N_1 = \cdots, \qquad M = M_1 = \cdots, \qquad \mathcal{P} = \mathcal{P}_1 = \cdots.$$

Since consequently $N \geq Q$, the assertion is proven.

2. $n = 2$, $m = 1$. Here every simultaneous covariant can be expressed rationally in terms of the co- and invariants

$$\begin{aligned} f &= a_0 x_1^2 + 2a_1 x_1 x_2 + a_2 x_2^2, \\ f_2 &= a_0 a_2 - a_1^2 = \mathcal{H}, \\ g &= b_0 x_1 + b_1 x_2, \\ s_1 &= (a_0 b_1 - a_1 b_0) x_1 + (a_1 b_1 - a_2 b_0) x_2 = \mathcal{I}. \end{aligned}$$

But we can easily form another invariant, namely, the resultant of the two forms f, g. The fact that resultants and discriminants are invariants will be proven later. But it follows already from the fact that if they are set equal to zero, they satisfy the necessary and sufficient condition for common, respectively, multiple linear factors, a property that is clearly invariant with respect to linear transformations. In our case, the resultant of f and g is:

$$\mathcal{R} = a_0 b_1^2 - 2a_1 b_1 b_0 + a_2 b_0^2.$$

We express this rationally in terms of f, f_2, g, s_1, according to our theorem. We have

$$\begin{aligned} \bar{\mathcal{R}} &= a_0 b_1^2 + a_2 b_0^2, \\ \bar{f}_2 &= a_0 a_2, \\ \bar{s}_1 &= a_0 b_1, \end{aligned}$$

and therefore

$$\mathcal{R} = f \frac{\mathcal{I}^2}{f^2} + \frac{\mathcal{H}}{f} g^2.$$

We have thus obtained the syzygy

$$\mathcal{R} f = \mathcal{I}^2 + \mathcal{H} g^2,$$

or

$$\mathcal{I}^2 = \mathcal{R} f - \mathcal{H} g^2;$$

and here, too, the square of the skew covariant is a polynomial function of the remaining ones, as was the case with all syzygies so far.

3. $n = 2$, $m = 2$. All in- and covariants are rational functions of

$$f = a_0 x_1^2 + \cdots,$$
$$g = b_0 x_1^2 + \cdots,$$
$$f_2 = a_0 a_2 - a_1^2 = \mathcal{H},$$
$$s_1 = (a_0 b_1 - a_1 b_0) x_1^2 + \cdots,$$
$$s_2 = a_0 b_2 - 2a_1 b_1 + a_2 b_0.$$

So, for instance, the invariant of g is

$$H = b_0 b_2 - b_1^2.$$

Here

$$\bar{f}_2 = a_0 a_2,$$
$$\bar{s}_1 = a_0 b_1,$$
$$\bar{s}_2 = a_0 b_2 + a_2 b_0,$$

whence

$$b_2 = \frac{\bar{s}_2 - a_2 b_0}{a_0}.$$

We have, therefore,

$$H = g \, \frac{s_2 - g \frac{\mathcal{H}}{f}}{f} - \frac{s_1^2}{f^2}$$

or

$$H = g \, \frac{s_2 f - g\mathcal{H}}{f^2} - \frac{s_1^2}{f^2},$$

so we obtain the syzygy

$$H f^2 = g s_2 f - g^2 \mathcal{H} - s_1^2$$

or

$$s_1^2 = g s_2 f - g^2 \mathcal{H} - f^2 H,$$

Again, the square of the skew covariant can be expressed as a polynomial of the others.

Lecture XXVII (June 25, 1897)

I.10 Covariants of covariants

If one fixes a covariant of one or more forms as base form, then one has the following:

Theorem *Covariants of covariants or simultaneous systems of covariants are again covariants.*

Remark Since invariants are special cases of covariants, the theorem holds also for invariants and simultaneous systems of invariants.

Namely, let

$$p_0 x_1^\pi + \cdots = \mathcal{P},$$
$$q_0 x_1^\kappa + \cdots = \mathcal{Q},$$
$$\cdots$$

be certain covariants of base forms; that is, if

$$p_0 = A(a, \ldots), \qquad p_1 = B(a, \ldots), \quad \ldots$$
$$p_0'' = A(a', \ldots), \qquad p_1'' = B(a', \ldots), \quad \ldots \quad ,$$

then we have

$$p_0'' x_1'^{\,\pi} + \cdots = \delta^{\lambda_\pi}(p_0 x_1^\pi + \cdots), \quad \ldots \quad .$$

On the other hand, the transformation sending a to a' was defined from the transformation sending x to x' via the identity; therefore after direct transformation

$$p_0' x_1'^{\,\pi} + \cdots = p_0 x_1^\pi + \cdots, \quad \ldots \quad .$$

Then obviously,

$$p_0'' = \delta^{\lambda_\pi} p_0', \quad \ldots \quad .$$

Now, let

$$\mathcal{S} = s_0 x_1^\sigma + \cdots$$

be a covariant of $\mathcal{P}$, $\mathcal{Q}, \ldots$; that is, let

$$s_0' x_1'^{\,\sigma} + \cdots = \delta^{\mu}(s_0 x_1^\sigma + \cdots),$$

where

$$s_0 = f(p_0, \ldots, q_0, \ldots), \quad \ldots$$
$$s_0' = f(p_0', \ldots, q_0', \ldots), \quad \ldots \quad .$$

But, according to the relations between p' and p'', etc., which we have just derived, we get

$$s'_0 = f\left(\frac{p''_0}{\delta\lambda_\pi}, \frac{p''_1}{\delta\lambda_\pi}, \ldots; \frac{q''_0}{\delta\lambda_\kappa}, \frac{q''_1}{\delta\lambda_\kappa}, \ldots; \ldots\right).$$

But since—as always—f is assumed to be homogeneous in the p as well as the q, one has:

$$s'_0 = \frac{1}{\delta^\lambda} f(p''_0, p''_1, \ldots; q''_0, q''_1, \ldots; \ldots).$$

Here, λ is the same for all s', since all the s have the same degree in the $p, \ldots$. Thus, if we set

$$s_0 = f(p_0, \ldots) = f\big(A(a, \ldots), \ldots\big) = \mathcal{F}(a, \ldots),$$

then

$$\delta^\lambda s'_0 = f(p''_0, \ldots) = f\big(A(a', \ldots), \ldots\big) = \mathcal{F}(a', \ldots)$$

and analogously for the other s'. But from this follows

$$\begin{aligned}\frac{1}{\delta^\lambda}\mathcal{F}(a, \ldots){x'_1}^\sigma &+ \frac{1}{\delta^\lambda}\mathcal{G}(a', \ldots){x'_1}^{\sigma-1}x'_2 + \ldots \\ &= \delta^\mu(s_0 x_1^\sigma + \cdots) \\ &= \delta^\mu\big(\mathcal{F}(a, \ldots)x_1^\sigma + \mathcal{G}(a, \ldots)x_1^{\sigma-1}x_2 + \cdots\big),\end{aligned}$$

or finally

$$\begin{aligned}\mathcal{F}(a', \ldots){x'_1}^\sigma &+ \mathcal{G}(a', \ldots){x'_1}^{\sigma-1}x'_2 + \cdots \\ &= \delta^{\lambda+\mu}\big(\mathcal{F}(a, \ldots)x_1^\sigma + \mathcal{G}(a, \ldots)x_1^{\sigma-1}x_2 + \cdots\big),\end{aligned}$$

which finishes the proof of the theorem.

This shows, for example, that transvections of transvections are again covariants, because transvections always remain homogeneous.

We illustrate this theorem with some examples.

1. A cubic base form:

$$f = a_0 x_1^3 + \cdots$$

The second transvection of the form f over itself is

$$\begin{aligned}\mathcal{H} &= (a_0a_2 - a_1^2)x_1^2 + (a_0a_3 - a_1a_2)x_1x_2 + (a_1a_3 - a_2^2)x_2^2 \\ &= \mathcal{H}_0x_1^2 + 2\mathcal{H}_1x_1x_2 + \mathcal{H}_2x_2^2.\end{aligned}$$

If one forms the transvection of $\mathcal{H}$ over itself twice, then one finds

$$\begin{aligned}\tfrac{1}{2}(\mathcal{H}, \mathcal{H})_2 &= \mathcal{H}_0\mathcal{H}_2 - \mathcal{H}_1^2 \\ &= (a_0a_2 - a_1^2)(a_1a_3 - a_2^2) - \tfrac{1}{4}(a_0a_3 - a_1a_2)^2,\end{aligned}$$

and this is indeed an invariant of f, namely, its discriminant.

2. If, in (1), one transvects f over $\mathcal{H}$ once, then one obtains

$$\begin{aligned}(f, \mathcal{H})_1 &= (a_0\mathcal{H}_1 - a_1\mathcal{H}_0)x_1^3 + \cdots \\ &= \tfrac{1}{2}(a_0^2a_3 - 3a_0a_1a_2 + 2a_1^3)x_1^3 + \cdots \\ &= \tfrac{1}{2}\mathcal{J};\end{aligned}$$

hence, up to a constant, this is the covariant of the cubic base form that we called $\mathcal{J}$ before (at the end of Lecture XX).

3. Let

$$\begin{aligned}f &= a_0x_1^2 + 2a_1x_1x_2 + a_2x_2^2, \\ g &= b_0x_1 + b_1x_2.\end{aligned}$$

Then

$$(f, g)_1 = (a_0b_1 - a_1b_0)x_1 + (a_1b_1 - a_2b_0)x_2 = \mathcal{I},$$

and furthermore

$$\begin{aligned}(g, \mathcal{I})_1 &= b_0(a_1b_1 - a_2b_0) - b_1(a_0b_1 - a_1b_0) \\ &= 2a_1b_0b_1 - a_2b_0^2 - a_0b_1^2;\end{aligned}$$

hence the resultant of a quadratic and a linear form is the functional determinant of the linear form and the functional determinant of the two base forms.

4. Let

$$f = a_0x_1^5 + \cdots$$

We form

$$\begin{aligned}\tfrac{1}{2}(f, f)_4 &= (a_0a_4 - 4a_1a_3 + 3a_2^2)x_1^2 + \cdots \\ &= C_0x_1^2 + 2C_1x_1x_2 + C_2x_2^2 = \mathcal{C}.\end{aligned}$$

Then

$$\tfrac{1}{2}(\mathcal{C}, \mathcal{C})_2 = C_0C_2 - C_1^2$$

is an invariant of degree four, and hence it must be the unique invariant of degree four of the form of order five, according to an earlier, not explicitly verified, remark.

Hence, our theorem shows that the formation of in- and covariants is, in some sense, a grouplike process in that one repeatedly obtains again in- and covariants.

We need to become acquainted, in this section, with two more processes which, when applied to an invariant system, preserve the invariant property. The first is the *polar process.* We have the following:

Theorem *The polars of a covariant possess the invariant property if the second sequence of variables y_1, y_2 is subjected to the same transformation as the first sequence of variables x_1, x_2.*

If C_x is an arbitrary expression that contains the variables x_1, x_2 homogeneously, then

$$\mathcal{P}_y(C_x) = y_1 \frac{\partial C}{\partial x_1} + y_2 \frac{\partial C}{\partial x_2}$$

is called the first polar of y with respect to C. In general, the polars are obtained if one expands

$$C(x_1 + \lambda y_1, x_2 + \lambda y_2)$$

in terms of powers of λ. From Taylor's Theorem one finds that

$$\begin{aligned} C(x_1 + \lambda y_1,\, x_2 + \lambda y_2) = C(x_1,\, x_2) &+ \frac{\lambda}{1!}\left(y_1 \frac{\partial C}{\partial x_1} + y_2 \frac{\partial C}{\partial x_2}\right) \\ &+ \frac{\lambda^2}{2!}\left(y_1^2 \frac{\partial^2 C}{\partial x_1^2} + 2y_1y_2 \frac{\partial C}{\partial x_1 \partial x_2} + y_2^2 \frac{\partial^2 C}{\partial x_2^2}\right) \\ &+ \quad \cdots \quad . \end{aligned}$$

The coefficients of $\frac{\lambda}{1!}$, $\frac{\lambda^2}{2!}, \ldots$ are then called the first, second, ... polars. One can easily see that the formation of the polars can be reduced to the formation of the first polar since the second polar is simply the first polar of the first polar, the third is the first polar of the second polar, etc.

Now, if C is a covariant, our theorem asserts that the polars of C are also covariants if the y are subjected to the same transformation formulas as the x. This is clear, because then apparently $x_1 + \lambda y_1$, $x_2 + \lambda y_2$ for arbitrary λ are also subjected to the same transformation formulas. Therefore we have

$$C(a_0', \ldots;\ x_1' + \lambda y_1', x_2' + \lambda y_2') = \delta^p C(a_0, \ldots;\ x_1 + \lambda y_1, x_2 + \lambda y_2),$$

and for each value of λ one must have:

$$\begin{aligned}
\mathcal{C}(a'_0, \ldots; x'_1, x'_2) &+ \frac{\lambda}{1!}\left(y'_1 \frac{\partial \mathcal{C}(a'; x')}{\partial x'_1} + y'_2 \frac{\partial \mathcal{C}(a'; x')}{\partial x'_2}\right) \\
&+ \frac{\lambda^2}{2!}\left(y_1'^2 \frac{\partial^2 \mathcal{C}(a'; x')}{\partial x_1'^2} + \cdots\right) + \cdots \\
&= \delta^p \left\{\mathcal{C}(a; x) + \frac{\lambda}{1!}\left(y_1 \frac{\partial \mathcal{C}(a; x)}{\partial x_1} + y_2 \frac{\partial \mathcal{C}(a; x)}{\partial x_2}\right)\right. \\
&\left. + \frac{\lambda^2}{2!}\left(y_1^2 \frac{\partial^2 \mathcal{C}(a; x)}{\partial x_1^2} + \cdots\right) + \cdots\right\}.
\end{aligned}$$

Since this equation is valid for all λ, the coefficients of equal powers of λ must possess the invariant property; so, indeed, every polar is a covariant. Invariants, covariants and polars of covariants are therefore again covariants of the base forms.

Lecture XXVIII (June 28, 1897)

Incidentally, we should remark that the weight of each polar of a covariant is equal to the weight of the covariant itself. If one applies the process of forming the kth polar to an expression, then the degree in the new variables y_1, y_2 is raised by k, and in those with respect to which one differentiates it is lowered by k. Besides, the number of groups of variables in $\mathcal{C}$ can be arbitrary, as long as they are all subjected to the same transformation formulas as the x and y. Considering the simplicity of the situation it seems unnecessary to provide examples; we only mention one briefly. The first polar of a form of order three

$$f = a_0 x_1^3 + 3a_1 x_1^2 x_2 + 3a_2 x_1 x_2^2 + a_3 x_2^3$$

is:

$$\begin{aligned}
\mathcal{P}_y(f_x^{(3)}) &= y_1(3a_0 x_1^2 + 6a_1 x_1 x_2 + 3a_2 x_2^2) \\
&\quad + y_2(3a_1 x_1^2 + 6a_2 x_1 x_2 + 3a_3 x_2^2) \\
&= 3(A_0 x_1^2 + 2A_1 x_1 x_2 + A_2 x_2^2),
\end{aligned}$$

where

$$\begin{aligned}
A_0 &= a_0 y_1 + a_1 y_2, \\
A_1 &= a_1 y_1 + a_2 y_2, \\
A_2 &= a_2 y_1 + a_3 y_2.
\end{aligned}$$

But since

$$A_0A_2 - A_1^2$$

is a covariant of $\mathcal{P}_y$, it is also a covariant of $f^{(3)}$, and indeed

$$\begin{aligned} A_0A_2 - A_1^2 &= (a_0y_1 + a_1y_2)(a_2y_1 + a_3y_2) - (a_1y_1 + a_2y_2)^2 \\ &= (a_0a_2 - a_1^2)y_1^2 + \cdots \end{aligned}$$

is the Hessian covariant of f, with y substituted for x.

A second process that preserves the invariant property is the so-called *Aronhold process*, which can be viewed as a generalization of the polar process. We have the following:

Theorem *If C is an invariant expression with respect to the simultaneous forms $f, g, h, \ldots$, and if furthermore $\mathcal{F}$ is a general form of the same degree as f, and if the coefficients*

$$a_0,\ a_1,\ a_2, \ldots$$

of f correspond to the coefficients

$$b_0,\ b_1,\ b_2, \ldots$$

of $\mathcal{F}$, then the function

$$b_0\frac{\partial C}{\partial a_0} + b_1\frac{\partial C}{\partial a_1} + b_2\frac{\partial C}{\partial a_2} + \cdots$$

also possesses the invariant property, provided the form $\mathcal{F}$ is added to the simultaneous system $f, g, \ldots$.

The proof of this theorem is similar to the one for polars. Replace the function f by $f + \lambda\mathcal{F}$, that is, a_0 by $a_0 + \lambda b_0$, a_1 by $a_1 + \lambda b_1, \ldots$. Then $a_i' + \lambda b_i'$ is related to $a_i + \lambda b_i$ through the same equations that relate a_i' to a_i. Since C is a covariant and no assumptions were made about the coefficients of f, we have

$$C(a_0' + \lambda b_0', \ldots; \ldots; x_1', x_2'; \ldots) = \delta^p C(a_0 + \lambda b_0, \ldots; \ldots; x_1, x_2; \ldots).$$

If one expands both sides in terms of powers of λ, then the right-hand side is identical to the left-hand side except that here the letters are primed. But since the equation is supposed to be valid identically in λ, the coefficients of equal powers of λ are equal, in particular those of λ, from which one sees that the function

$$b_0\frac{\partial C}{\partial a_0} + b_1\frac{\partial C}{\partial a_1} + \cdots$$

above is indeed a covariant whose weight is the same as that of C.

For instance, from the invariant of the form of order four

$$a_0a_4 - 4a_1a_3 + 3a_2^2$$

one can derive in this way the simultaneous invariant of two forms of order four:

$$b_0a_4 - 4b_1a_3 + 6b_2a_2 - 4b_3a_1 + b_4a_0,$$

and this is indeed the fourth transvection of the two forms one over the other.

That the polar process is a special case of the Aronhold process just discussed can be seen as follows. Let C be a covariant containing x_1 and x_2; if we replace x_1 and x_2 by $-b_1$ and b_0, respectively, where $b_0x_1 + b_1x_2$ is a linear form, then we obtain a simultaneous invariant. If we add a linear form $a_0x_1 + a_1x_2$, then

$$a_0 \frac{\partial C(-b_1, b_0)}{\partial b_0} + a_1 \frac{\partial C(-b_1, b_0)}{\partial b_1}$$

is also an invariant. We rewrite this expression as

$$a_0 \frac{\partial C(-b_1, b_0)}{\partial b_0} - a_1 \frac{\partial C(-b_1, b_0)}{\partial(-b_1)}.$$

From this we obtain again a covariant if we replace a_0, a_1 by y_2, $-y_1$, and b_0, b_1 by x_2, $-x_1$, respectively; that is, the polar

$$y_1 \frac{\partial C}{\partial x_1} + y_2 \frac{\partial C}{\partial x_2}$$

is indeed a covariant.

I.11 The invariants and covariants as functions of the roots

Until now we have almost exclusively considered the invariants and covariants as functions of the coefficients of the base forms and the variables; in the three following sections we need to introduce three new ways of representing them.

We begin by considering at first only one base form:

$$f = a_0x_1^n + \binom{n}{1}a_1x_1^{n-1}x_2 + \cdots + a_nx_2^n.$$

According to the Fundamental Theorem of Algebra one can decompose f into exactly n linear forms, so one has

$$f = a_0(x_1 - \omega_1x_2)(x_1 - \omega_2x_2)\cdots(x_1 - \omega_nx_2),$$

where $\omega_1, \omega_2, \ldots, \omega_n$ are the n roots of the equation $f = 0$ in x_1/x_2. The quotients a_i/a_0 are symmetric polynomial functions of the ω. A homogeneous in- or covariant can be written in the form

$$\mathcal{C} = a_0^N \mathcal{C}\left(1, \frac{a_1}{a_0}, \frac{a_2}{a_0}, \ldots, \frac{a_n}{a_0}; x_1, x_2\right),$$

thus it can also be written as a symmetric polynomial function of the roots ω, if one adds a suitable power of a_0 as an additional factor. Suppose given an invariant $\mathcal{I}(a_0, a_1, \ldots, a_n)$ we write

$$\mathcal{I}(a_0, a_1, \ldots, a_n) = a_0^N \mathcal{G}(\omega_1, \omega_2, \ldots, \omega_n)$$

and ask: How can we determine from $\mathcal{G}$ whether $\mathcal{G}$ is an invariant?

To this aim, we observe how a linear transformation transforms the roots. Let

$$\begin{aligned} x_1 &= \alpha_{11}x_1' + \alpha_{12}x_2', \\ x_2 &= \alpha_{21}x_1' + \alpha_{22}x_2'; \end{aligned}$$

hence

$$\begin{aligned} \delta x_1' &= \alpha_{22}x_1 - \alpha_{12}x_2, \\ \delta x_2' &= -\alpha_{21}x_1 + \alpha_{11}x_2. \end{aligned}$$

Under this transformation, f becomes

$$f = f' = a_0'(x_1' - \omega_1' x_2')\,(x_1' - \omega_2' x_2') \cdots (x_1' - \omega_n' x_2').$$

Each linear factor is transformed into another linear factor, and we suppose that the ω_i and the ω_i' are related in this fashion. Then

$$\begin{aligned} x_1 - \omega_i x_2 &= \alpha_{11}x_1' + \alpha_{12}x_2' - \omega_i(\alpha_{21}x_1' + \alpha_{22}x_2') \\ &= c_i(x_1' - \omega_i' x_2'); \end{aligned}$$

therefore

$$\begin{aligned} \alpha_{11} - \omega_i\alpha_{21} &= c_i, \\ \alpha_{12} - \omega_i\alpha_{22} &= -c_i\omega_i', \end{aligned}$$

and consequently

$$\omega_i' = \frac{\alpha_{22}\omega_i - \alpha_{12}}{\alpha_{11} - \omega_i\alpha_{21}}.$$

Using this fact, we now prove the following:

Theorem *The expression*

$$\mathcal{I} = a_0^w \sum \prod (\omega_i - \omega_k)$$

is an invariant of the form f, provided that each of the products $\prod$ is a homogeneous function of the differences, say of degree p, that each of the products contains each root ω the same number of times, say w, and that the whole sum is a symmetric function of the roots ω.

Lecture XXIX (June 29, 1897)

This theorem is easy to prove. Namely, we have

$$\omega_i' - \omega_k' = \frac{\alpha_{22}\omega_i - \alpha_{12}}{\alpha_{11} - \omega_i\alpha_{21}} - \frac{\alpha_{22}\omega_k - \alpha_{12}}{\alpha_{11} - \omega_k\alpha_{21}},$$

or

$$\begin{aligned}\omega_i' - \omega_k' &= \frac{(\alpha_{22}\omega_i - \alpha_{12})(\alpha_{11} - \omega_k\alpha_{21}) - (\alpha_{22}\omega_k - \alpha_{12})(\alpha_{11} - \omega_i\alpha_{21})}{(\alpha_{11} - \omega_i\alpha_{21})(\alpha_{11} - \omega_k\alpha_{21})} \\ &= \frac{\omega_i(\alpha_{22}\alpha_{11} - \alpha_{12}\alpha_{21}) - \omega_k(\alpha_{11}\alpha_{22} - \alpha_{12}\alpha_{21})}{(\alpha_{11} - \omega_i\alpha_{21})(a_{11} - \omega_k\alpha_{21})} \\ &= \delta\,\frac{\omega_i - \omega_k}{(\alpha_{11} - \omega_i\alpha_{21})(\alpha_{11} - \omega_k\alpha_{21})}.\end{aligned}$$

The product $\prod$ contains p such differences. Thus, if we form

$$\mathcal{I}' = a_0'^{\,w}\sum\prod(\omega_i' - \omega_k'),$$

for which we want to prove that

$$\mathcal{I}' = \delta^p\mathcal{I},$$

then $\prod$ contains the factor δ^p; hence so does $\mathcal{I}'$ because of homogeneity. Therefore, we have

$$\mathcal{I}' = a_0'^{\,w}\delta^p\sum\prod\frac{\omega_i - \omega_k}{(\alpha_{11} - \omega_i\alpha_{21})(\alpha_{11} - \omega_k\alpha_{21})}.$$

But, furthermore, $\prod$ contains each ω_i w times; so it follows that

$$\begin{aligned}\mathcal{I}' = a_0'^{\,w}\delta^p\sum &\frac{1}{\left[(\alpha_{11} - \omega_1\alpha_{21})(\alpha_{11} - \omega_2\alpha_{21})\cdots(\alpha_{11} - \omega_n\alpha_{21})\right]^w} \\ &\times\prod(\omega_i - \omega_k).\end{aligned}$$

But we also have (Lecture XXVIII)

$$\begin{aligned}a_0' &= a_0c_1c_2\cdots c_n \\ &= a_0(\alpha_{11} - \omega_1\alpha_{21})(\alpha_{11} - \omega_2\alpha_{21})\cdots(\alpha_{11} - \omega_n\alpha_{21}),\end{aligned}$$

so it follows that

$$\mathcal{I}' = \delta^p a_0^w \sum \prod (\omega_i - \omega_k),$$

which completes the proof of the theorem.

It also follows that: *The weight of the invariant $\mathcal{I}$ is p, the degree is w.*

We give some examples. The discriminant of a form of order n is

$$\mathcal{D} = a_0^{2(n-1)} \prod (\omega_i - \omega_k)^2.$$

There are $n(n-1)$ factors altogether; thus the relation $ng - 2p = 0$ is satisfied, that is,

$$n2(n-1) - 2n(n-1) = 0,$$

as it has to be. The discriminant is an invariant.

Likewise, for $n = 4$,

$$a_0^2 \sum (\omega_1 - \omega_2)^2 (\omega_3 - \omega_4)^2$$

has to be an invariant of degree two and weight four. Since there is only one such, we conclude that this expression has to be equal to

$$a_0^2 \sum (\omega_1 - \omega_2)^2 (\omega_3 - \omega_4)^2 = c(a_0 a_4 - 4a_1 a_3 + 3a_2^2).$$

For $n = 4$, the expression

$$a_0^3 \sum (\omega_1 - \omega_2)^2 (\omega_3 - \omega_4)^2 (\omega_1 - \omega_3)(\omega_2 - \omega_4)$$

is also an invariant of degree three and weight six. Since there is only one such, namely, $j = a_0 a_2 a_4 + 2a_1 a_2 a_3 - a_2^3 - a_0 a_3^2 - a_1^2 a_4$, it follows that this expression has to be equal to $c_1 \cdot j$.

On the other hand, for $n = 4$ we also have the invariant

$$a_0^3 \prod \{(\omega_1 - \omega_3)(\omega_2 - \omega_4) + (\omega_2 - \omega_3)(\omega_1 - \omega_4)\}$$

of degree three and weight six. Hence, this expression must be essentially identical to the previous one, namely, equal to $c_2 \cdot j$. From this, one recognizes at the same time the significance of $j = 0$. In this case, one of the three factors is equal to zero, say

$$\frac{\omega_1 - \omega_3}{\omega_2 - \omega_3} = -\frac{\omega_1 - \omega_4}{\omega_2 - \omega_4}$$

or

$$\frac{\omega_1 - \omega_3}{\omega_2 - \omega_3} : \frac{\omega_1 - \omega_4}{\omega_2 - \omega_4} = -1,$$

and if one makes an analogous calculation for the two other factors, then one sees that $j = 0$ is the necessary and sufficient condition for the four roots of f to have harmonic position* in some combination.

The theorem we have proven also has a converse.

Theorem *Each invariant of the form f is a symmetric function of the roots that depends homogeneously only on the differences and that contains each root in each term (written as a function of the differences of the roots) to the same degree.*

We omit the proof of this theorem. Furthermore, the theorem can easily be generalized to simultaneous invariants. If

$$f = a_0 x_1^n + \binom{n}{1} a_1 x_1^{n-1} x_2 + \cdots = a_0(x_1 - \omega_1 x_2) \cdots (x_1 - \omega_n x_2),$$

$$g = b_0 x_1^m + \binom{m}{1} b_1 x_1^{m-1} x_2 + \cdots = b_0(x_1 - \rho_1 x_2) \cdots (x_1 - \rho_m x_2),$$

then we have the following:

Theorem *The expression*

$$\mathcal{I} = a_0^w b_0^u \sum \prod (\omega_i - \omega_k) \prod (\rho_i - \rho_k) \prod (\omega_i - \rho_k)$$

is an invariant if the sum is a symmetric function of the ω as well as the ρ, which is homogeneous in all root differences, in such a way that in each term of the sum each ω appears equally often, namely, w times, and each ρ appears equally often, namely, u times.

This theorem too has a converse, and can also be generalized to arbitrarily many forms.

The resultant of the two forms f and g provides an example. Recall that it is equal to

$$a_0^w b_0^u \prod (\omega_i - \rho_k) = a_0^m b_0^n \prod (\omega_i - \rho_k),$$

so it is indeed an invariant. The condition $nr + ms = 2p$ is satisfied, since $p = mn$, $w = m$, and $u = n$.

For $n = 2$, $m = 2$,

$$a_0 b_0 \{(\omega_1 - \rho_1)(\omega_2 - \rho_2) + (\omega_2 - \rho_1)(\omega_1 - \rho_2)\}$$

is an invariant; indeed, it is equal to

$$2(a_0 b_2 - 2a_1 b_1 + a_2 b_0).$$

* Four points are in *harmonic position* if the cross ratios of their coordinates are equal to -1.

Thus, we have at the same time implicitly treated the case of covariants.

I.12 The invariants and covariants as functions of the one-sided derivatives

Every binary form is essentially identical to a polynomial function in one variable. This polynomial is obtained by setting $x_1 = x$, $x_2 = 1$, and we will write:

$$f = f^{(n)}(x, 1) = a_0 x^n + \binom{n}{1} a_1 x^{n-1} + \cdots + a_n.$$

Furthermore, we want to introduce the notation

$$f_i = \frac{(n-i)!}{n!} \frac{d^i f(x, 1)}{dx^i},$$

so that f_i is in essence the ith derivative with respect to the one variable (the ith one-sided derivative). We then have the following:

Theorem *Every covariant C of the form f can be directly expressed as a polynomial function of the one-sided derivatives f_i; one obtains this expression by substituting f_i for a_i in the source of the covariant.*

Of course, this theorem is also valid for invariants. We omit the proof. In any case, it demonstrates once again that there is a connection between invariant theory and differential equations. In fact, it is possible, under certain circumstances, to integrate differential equations invariant theoretically. Since the intrinsic properties of forms are given by the vanishing of in- and covariants, we see that we can also characterize them through differential equations, which one knows as soon as the in- and covariants are known.

Lecture XXX (July 1, 1897)

Representation of the in- and covariants in terms of one-sided derivatives is of great importance for many problems. We shall only mention one, which was already treated by Clebsch. Suppose that one or more forms, whose coefficients are not arbitrary anymore, but have been specialized in some way, satisfy certain conditions. The problem is to determine necessary and sufficient conditions in such a way that forms that satisfy them can be transformed linearly into the original ones; that is, one should find the "invariant criteria" for the specialization.

We can treat this problem with our representation and carry it out completely for spherical functions, for instance.

We illustrate the representation of in- and covariants in terms of one-sided derivatives with some examples. The Hessian covariant is

$$\mathcal{H} = (a_0 a_2 - a_1^2) x_1^{2n-4} + \cdots,$$

so we have

$$\begin{aligned}\mathcal{H} &= f_0 f_2 - f_1^2 \\ &= f\,\frac{(n-2)!}{n!} f'' - \left[\frac{(n-1)!}{n!}\right]^2 f'^2\end{aligned}$$

or

$$\mathcal{H} = \frac{1}{n^2(n-1)}\left(nff'' - (n-1)f'^2\right).$$

Likewise, the Jacobian covariant is

$$\begin{aligned}(f,g)_1 &= (a_0 b_1 - a_1 b_0) x_1^{n+m-2} + \cdots \\ &= f_0 g_1 - g_0 f_1\end{aligned}$$

or

$$(f,g)_1 = \frac{1}{m} f g' - \frac{1}{n} f' g.$$

In general, the pth transvection of f over g is

$$(f,g)_p = f_0 g_p - \binom{p}{1} f_1 g_{p-1} + \binom{p}{2} f_2 g_{p-2} - \cdots + (-1)^p f_p g_0.$$

Of course, for invariants, the variables have to disappear from the representation when one substitutes explicity expressions for the one-sided derivatives. For example, for $n = 2$, the Hessian covariant is

$$\begin{aligned}\mathcal{H} &= \tfrac{1}{4}(2ff'' - f'^2) \\ &= \tfrac{1}{4}\left(2(a_0x^2 + 2a_1x + a_2)2a_0 - (2a_0x + 2a_1)^2\right) \\ &= a_0^2x^2 + 2a_0a_1x + a_0a_2 - a_0^2x^2 - 2a_0a_1x - a_1^2 \\ &= a_0a_2 - a_1^2.\end{aligned}$$

One can easily construct more such examples.

Remark The main theorem of this section has been stated and used, especially for spherical functions, by Prof. Hilbert in his inaugural dissertation: "Über die Invarianten Eigenschaften Spezieller Binärer Formen,

Insbesondere der Kugelfunktionen," Königsberg, Prussia, 1885. The results herein are generalized to the terminating hypergeometric series

$$\mathcal{F}(\alpha, \beta, \gamma, x) = 1 + \frac{\alpha\beta}{1\cdot\gamma}x + \frac{\alpha(\alpha+1)\beta(\beta+1)}{1\cdot 2\cdot\gamma(\gamma+1)}x^2 + \cdots,$$

in the case that $\beta = -n$ is a negative integer, in the treatise: "Über eine Darstellungsweise der Invarianten Gebilde im Binären Formengebiete" by D. Hilbert, *Math. Ann.* **30**. There it is shown that for the latter, one can use that theorem to determine very simple and interesting invariant conditions.

I.13 The symbolic representation of invariants and covariants

After we have seen two new representations of in- and covariants in the previous two sections, we now finally come to a representation that is entirely different from the previous ones. The *symbolic representation* of invariants and covariants was the one used almost exclusively in earlier textbooks, notably in England, America, and Italy.* Even Cayley used a symbolism for invariant-theoretical investigations. The method, however, which will concern us subsequently, which allows simple symbolic computations often convenient for the derivation of the in- and covariants, originates with Clebsch, and was practiced after him, notably by Gordan. We begin with a form and consider its invariants; the consideration of covariants and simultaneous in- and covariants then presents no additional difficulties; for them we will only state the appropriate theorems, however.

Let

$$f = a_0 x_1^n + \binom{n}{1} a_1 x_1^{n-1} x_2 + \cdots + a_n x_2^n$$

be the given base form, and let

$$\mathcal{I} = \sum Z_{\nu_0 \cdots \nu_n} a_0^{\nu_0} a_1^{\nu_1} \cdots a_n^{\nu_n}$$

be one of its invariants. Instead of $a_0^{\nu_0}$ we write explicitly $a_0 a_0 \cdots a_0$ (ν_0 times), etc. Then the invariant is

$$\mathcal{I} = \sum Z_{\mu_0\mu_1\mu_2\cdots} a_{\mu_0} a_{\mu_1} a_{\mu_2} a_{\mu_3} \cdots,$$

where the number of the occurring a is obviously the degree; furthermore, the sum of the subscripts is the weight of each term, and hence

* See Kung and Rota (1984) for a modern treatment of the symbolic method for invariants of binary forms.

also of the invariant. Now we substitute a b for the second a, etc., and write

$$\mathcal{I} = \left[\sum Z_\mu a_{\mu_1} b_{\mu_2} c_{\mu_3} \cdots\right]_{a=b=c=\cdots=a}.$$

in such a way that each term of the invariant contains exactly *one* a, *one* $b, \ldots$. But then we can, without ambiguity, write exponents instead of subscripts; that is,

$$\mathcal{I} = \left[\left[\sum Z_\mu a^{\mu_1} b^{\mu_2} c^{\mu_3} \cdots\right]\right]_{\downarrow a=b=c=\cdots=a}.$$

Because from this symbolic expression we obtain completely and uniquely the explicit expression if we write subscripts instead of exponents (which is indicated via the arrow $\downarrow$), and then set $a = b = c = \cdots$. The new variables $a, b, c, \ldots$ are called the *symbols*, which give the representation. One uses as many symbols as the degree of the invariant.

This representation, however, would complicate matters rather than simplify them if one could not write the resulting expression more simply, provided one thinks of $a^{\mu_1}, \ldots$ as real powers. But for this it is necessary that one introduces the symbols $a, b, c, \ldots$ into the individual terms appropriately; because, while it is initially arbitrary for which a in each term one substitutes b, and for which a one substitutes c, ..., it is not arbitrary anymore if one wants to achieve a simple representation, which is always possible. At the very least, the arbitrariness is always very limited.

To give an example of this, consider the invariant of the quadratic form. Thus, we write

$$\begin{aligned}
\mathcal{I} &= a_0 a_2 - a_1^2 \\
&= \tfrac{1}{2}(a_0 a_2 - 2a_1^2 + a_2 a_0) \\
&= \tfrac{1}{2}(a_0 b_2 - 2a_1 b_1 + a_2 b_0)_{b=a} \\
&= \tfrac{1}{2}(b^2 - 2ab + a^2)_{b=a}
\end{aligned}$$

or

$$\mathcal{I} = \tfrac{1}{2}(b-a)^2.$$

From this expression one obtains the explicit representation if one *first* expands, replaces exponents by subscripts, and finally substitutes $a = b$. In homogeneous expressions, one has to insert a^0 or b^0 ..., depending on the missing symbols.

When can one tell from looking at such a symbolic expression whether it is an invariant or not? We will derive a sufficient condition for this. If an expression $\mathcal{I}$ is to be an invariant, then it has to satisfy the differential equation

$$\mathcal{D}\mathcal{I} = 0.$$

If we use the symbolic representation, then this becomes

$$\mathcal{D}\mathcal{I} = \mathcal{D}_a\big[[\ldots]\big]_{b=\cdots=a} + \mathcal{D}_b\big[[\ldots]\big]_{b=\cdots=a} + \cdots.$$

But we have

$$\mathcal{D}_a = a_0 \frac{\partial}{\partial a_1} + 2a_1 \frac{\partial}{\partial a_2} + \cdots + na_{n-1} \frac{\partial}{\partial a_n}.$$

In the symbolic representation, the expression

$$ia^{i-1} \frac{\partial}{\partial(a^i)} = \frac{\partial(a^i)}{\partial a} \frac{\partial}{\partial(a^i)}$$

replaces

$$ia_{i-1} \frac{\partial}{\partial a_i};$$

therefore

$$\mathcal{D}_a = \frac{\partial a}{\partial a} \frac{\partial}{\partial a} + \frac{\partial(a^2)}{\partial a} \frac{\partial}{\partial(a^2)} + \frac{\partial(a^3)}{\partial a} \frac{\partial}{\partial(a^3)} + \cdots + \frac{\partial(a^n)}{\partial a} \frac{\partial}{\partial(a^n)},$$

where all powers of a are to be considered independent for the moment.

One therefore obtains:

$$\mathcal{D}_a = n \cdot \frac{\partial}{\partial a},$$

where now the powers are to be considered as such, that is, as depending on a. An invariant in its symbolic form must therefore satisfy the differential equation

$$\frac{\partial\big[[\ldots]\big]}{\partial a} + \frac{\partial\big[[\ldots]\big]}{\partial b} + \frac{\partial\big[[\ldots]\big]}{\partial c} + \cdots = 0.$$

So far we have thus found:

Theorem *The expression*

$$\mathcal{I} = \sum Za^{\mu_1} b^{\mu_2} c^{\mu_3} \cdots$$

is the symbolic representation of an invariant if and only if each term contains equally many, say g, symbols, if furthermore each term has the same total degree, say p, where $ng = 2p$, and if it satisfies the differential equation

$$\frac{\partial\mathcal{I}}{\partial a} + \frac{\partial\mathcal{I}}{\partial b} + \frac{\partial\mathcal{I}}{\partial c} + \cdots = 0.$$

This invariant, in its nonsymbolic representation, is of degree g and weight p.

Lecture XXXI (July 2, 1897)

It is now very easy to give a polynomial function of the symbols $a, b, c, \ldots$ that satisfies these conditions, namely, an arbitrary polynomial function of the differences $b-a$, $c-a, \ldots$, as long as the homogeneity conditions are observed. Because of the importance attached to these differences, one also writes them in the form:

$$b-a=(ab),$$
$$c-a=(ac),$$

etc. A polynomial function of these differences is written as

$$\mathcal{S}=\sum Z(ab)^{\sigma}(ac)^{\epsilon}(bc)^{\rho}(bd)^{\kappa}\cdots(al)^{\lambda}.$$

This obviously satisfies the differential equation

$$\frac{\partial \mathcal{S}}{\partial a}+\frac{\partial \mathcal{S}}{\partial b}+\cdots=0,$$

since each factor of each term satisfies it:

$$\frac{\partial (ab)^{\sigma}}{\partial a}+\frac{\partial (ab)^{\sigma}}{\partial b}+\cdots=-\sigma(ab)^{\sigma-1}+\sigma(ab)^{\sigma-1}+0=0.$$

Furthermore, the sum of the exponents

$$\sigma+\epsilon+\rho+\cdots+\lambda=p$$

must be constant in each term of the sum. Now suppose the symbol a occurs α times, the symbol b occurs β times, ... in each term of the sum; then apparently

$$\alpha+\beta+\cdots=2(\sigma+\epsilon+\rho+\cdots)=2p,$$

and therefore the condition

$$\alpha+\beta+\gamma+\cdots=ng$$

has to be satisfied also. Now, we necessarily have

$$\alpha\leq n,\ \beta\leq n,\ \gamma\leq n,\ldots,$$

since only the symbols $a^0, a^1, \ldots, a^n$; $b^0, b^1, \ldots, b^n$; ... make sense. But the number of symbols that occur in each term is equal to g, so the number of quantities $\alpha, \beta, \gamma, \ldots$ also has to equal g. If even only one of the quantities $\alpha, \beta, \gamma, \ldots$ were less than n, then its sum could not be equal to ng, which is required. Hence, it follows that

$$\alpha = n, \ \beta = n, \ \gamma = n, \ldots$$

has to hold. And thus we have obtained the following:

Theorem 1 *The symbolic expression*

$$\mathcal{S} = \sum Z(ab)^\sigma (ac)^\tau (bc)^\rho \cdots (al)^\lambda$$

is an invariant of degree g and weight p, if p is the sum of the exponents $\sigma + \tau + \cdots + \lambda$, common to all terms, and if, furthermore, each of the g symbols occurs exactly n times in each individual term.

Incidentally, the converse holds also.

Theorem *Each invariant can be represented symbolically as a function of the differences.*

We shall give a brief sketch of the proof further below.

One should remark that writing the coefficients symbolically amounts to writing the base form in the form

$$f = (x_1 + ax_2)^n = (x_1 + bx_2)^n = \ldots,$$

where the $a^i = a_i$, $b^i = a_i$, etc. have the same meaning and are only introduced together to preserve uniqueness.

We illustrate the above with some examples. Clearly,

$$(ab)^n = \mathcal{I}$$

is an invariant of the form of order n. Indeed,

$$\begin{aligned}
(ab)^n &= (b-a)^n \\
&= b^n - \binom{n}{1} b^{n-1} a + \binom{n}{2} b^{n-2} a^2 - \cdots \\
&= a_0 a_n - \binom{n}{1} a_1 a_{n-1} + \binom{n}{2} a_2 a_{n-2} - \cdots + (-1)^n a_n a_0.
\end{aligned}$$

If n is odd, then this expression vanishes identically. But if n is even, then this is exactly the nth transvection of the form f over itself.

In particular, if for example $n = 4$, then

$$(ab)^4 = b^4 - 4b^3a + 6b^2a^2 - 4ba^3 + a^4$$

or

$$\begin{aligned}(ab)^4 &= a_0a_4 - 4a_1a_3 + 6a_2^2 - 4a_3a_1 + a_4a_0 \\ &= 2(a_0a_4 - 4a_1a_3 + 3a_2^2) \\ &= 2i.\end{aligned}$$

Similarly, one sees that

$$(ab)^2(ac)^2(bc)^2 = 6j.$$

But, in order to give an example of how one can represent explicitly given invariants symbolically, as well as at the same time give an example for such symbolic calculations, we want, conversely, to try to express j in symbolic form. We have

$$j = \begin{vmatrix} a_0 & a_1 & a_2 \\ a_1 & a_2 & a_3 \\ a_2 & a_3 & a_4 \end{vmatrix}$$

$$= \begin{vmatrix} a_0 & a_1 & a_2 \\ b_1 & b_2 & b_3 \\ c_2 & c_3 & c_4 \end{vmatrix},$$

whence

$$\begin{aligned} j = \begin{vmatrix} 1 & a & a^2 \\ b & b^2 & b^3 \\ c^2 & c^3 & c^4 \end{vmatrix} &= bc^2 \begin{vmatrix} 1 & a & a^2 \\ 1 & b & b^2 \\ 1 & c & c^2 \end{vmatrix} \\ &= bc^2(ab)\,(ac)\,(bc). \end{aligned}$$

Needless to say, it is irrelevant in which row of the determinant we initially place the a, in which the b, and in which the c, since all three symbols have the same meaning; in general, we have: *"One can interchange symbols that have the same value,"* which is obvious. If we now interchange a and b in the last expression for j, then one finds that

$$j = -ac^2(ab)\,(ac)\,(bc),$$

and addition of both formulas gives

$$2j = c^2(ab)\,(ab)\,(ac)\,(bc).$$

If we now furthermore interchange c with a, and c with b, then one finds that

$$2j = a^2(bc)\,(bc)\,(ac)\,(ab),$$
$$2j = -b^2(ac)\,(ac)\,(ab)\,(bc).$$

Addition of these three formulas gives

$$6j = (ab)\,(ac)\,(bc)\begin{vmatrix} 1 & a & a^2 \\ 1 & b & b^2 \\ 1 & c & c^2 \end{vmatrix},$$

hence

$$6j = (ab)^2(ac)^2(bc)^2,$$

as we just claimed.

For covariants and simultaneous in- and covariants we have theorems analogous to those for the invariants of a single form. We list the theorems without proofs and remark that the proofs for the simultaneous invariants can be carried out just like those for the invariants of a single form, and that then the proofs for the covariants result from the fact that each covariant has to become an invariant if one sets $x_1 = -b_1$, $x_2 = b_0$, where b_0, b_1 are the coefficients of a linear form, and conversely. If we use the abbreviation

$$x_1 + ax_2 = a_x,$$

so that the base form is $f = a_x^n = b_x^n = \ldots$, then the theorems in question are as follows.

Theorem 2 *The symbolic expression*

$$\sum Z(ab)^\sigma(ac)^\tau(bc)^\kappa \cdots a_x^\xi b_x^\eta \cdots$$

is a covariant of degree g, order m, and weight p, if p is the sum of the exponents $\sigma + \tau + \kappa + \cdots$ common to all summands, and m is the common sum of exponents $\xi + \eta + \cdots$, and if furthermore each of the g symbols a, b, $c, \ldots$ occurs n times in each summand.

So, for example, we have

$$(f,\, f)_i = (ab)^i a_x^{n-i} b_x^{n-i},$$

the ith transvection of the form f over itself; in particular we have

$$\mathcal{H} = (ab)^2 a_x^{n-2} b_x^{n-2},$$

the Hessian covariant of the form $f = a_x^n = b_x^n$.

Theorem 3 *The symbolic expression*

$$\sum Z(ab)^{\sigma}(ac)^{\tau}\cdots(aa')^{\mu}(ba')^{\xi}\cdots(a'b')^{\kappa}(a'c')^{\lambda}\cdots$$

is a simultaneous invariant of degree g *in the coefficients of* $f = a_x^n = b_x^n = \cdots$, *and of degree* g' *in the coefficients of* $f' = {a'_x}^{n'} = {b'_x}^{n'} = \cdots$, *etc., and of weight* p, *if* p *is the sum of the exponents, common to all summands, and if, moreover, each of the* g *symbols* $a, b, c, \ldots$ *occurs* n *times in each summand, whereas each of the* g' *symbols* $a', b', c', \ldots$ *occurs* n' *times in each summand, etc.*

An analogous theorem holds for simultaneous covariants. The converse of these theorems holds also. One can represent each invariant of one or more forms symbolically as a polynomial function of the differences (ab), $(ac), \ldots$, and one can, similarly, express each covariant of one or more forms in such a way that it appears symbolically as a polynomial function of the types (ab) and a_x.

The idea for a proof of this converse is based on the following: Suppose we are given an invariant

$$\mathcal{I} = \mathcal{I}(a_0, \ldots, a_n; a'_0, \ldots, a'_n; \ldots).$$

Then

$$b_0 \frac{\partial \mathcal{I}}{\partial a_0} + b_1 \frac{\partial \mathcal{I}}{\partial a_1} + \cdots + b_n \frac{\partial \mathcal{I}}{\partial a_n}$$

is also an invariant, which is linear in the b, and whose degree in a is lowered by one. Furthermore, it is transformed into the original one if one identically equates the new form that was added with the form with the coefficients a, that is, if one sets $a_0 = b_0$, $a_1 = b_1, \ldots$. By adding more forms, one can clearly obtain an invariant that contains the coefficients of all its base forms linearly and is transformed, up to a constant, into the invariant $\mathcal{I}$ if one sets the added forms identically equal to the ones to which they were added. One can now write exponents instead of subscripts without creating ambiguities, and obtains a symbolic representation like the one at the beginning of this section. But if one considers this representation as an actual invariant, that is, if one considers the exponents as real exponents, then the resulting expression must be a simultaneous invariant of only linear forms. But each simultaneous invariant of linear forms is a polynomial function of the differences (ab), $(ac), \ldots$, as one can prove, in somewhat complicated fashion, inductively for simultaneous forms (see, for instance, the proof in Clebsch and Lindemann, *Geometrie*, vol. I, p. 184 ff.). The symbolic expression must, therefore, always be expressible as a polynomial function of the differences—which also proves the converse.

Thus, if one wants to form all possible invariants and covariants, one only needs to form all possible polynomial functions of the types (ab) and a_x, while observing the side conditions. It can thereby happen that an expression takes on different meanings depending on which base form one considers; for instance, every symbolic expression represents a simultaneous invariant of linear forms. Conversely, it can also happen that a given invariant has several different symbolic representations. When making symbolic calculations, one of course chooses the simplest one.

Final remarks to Part I

If we now recapitulate briefly what we have discussed so far, then it consisted mostly of three things—disregarding the enumeration calculus that we employed. After defining invariants and covariants, we first derived necessary and sufficient conditions for them. There, we used the essential fact that each linear transformation can be reduced to three especially simple linear transformations. We restricted ourselves to binary forms, but it is evident that this restriction is not essential, that indeed these reflections can be extended easily to forms with arbitrarily many variables. Second, we constructed a system of covariants for which we could prove that all other possible covariants can be expressed rationally through these. This theorem, which can also be generalized to forms with arbitrarily many variables, proved to be particularly fruitful for the construction of the covariants and the syzygies existing between them. Third, we gave three new representations of covariants, of which the symbolic representation led to a simple way of constructing covariants, and which has the advantage of being directly applicable to forms with several variables. We can perhaps characterize the three parts in the following schematic way:

1. Definitions and concepts. Basic properties.
2. Fundamental Theorem. Examples and corollaries to it.
3. Methods of representation.

The rational representation of all covariants through a finite system of them directly raises the question whether, for a given system of simultaneous forms, one can construct a finite system of in- and covariants through which all the others can be expressed as *polynomials*. We now have to consider this question, which is substantially deeper than the one regarding rational representation, and which has an affirmative answer. To answer it we need to make observations that are completely different from the ones so far. The theorem about the finiteness of a full invariant system—first proven in general by Hilbert—is of fundamental importance for higher invariant theory. All further observations are based on it; it forms the foundation for the second part of the theory as well.

II
The theory of invariant fields

Lecture XXXII (July 5, 1897)

II.1 Proof of the finiteness of the full invariant system via representation by root differences

The invariants of a simultaneous system of base forms constitute an integral domain, insofar as a polynomial function of invariants or covariants is again an invariant or covariant, as long as homogeneity—rather only the power of the substitution determinant—is preserved. Thus, there arises the following question: Does this integral domain have a finite basis? That is: Is there a finite system of invariants through which all others can be expressed as polynomial functions? And one has to ask furthermore: If there is such a "finite full invariant system" (respectively, system of forms), how many mutually independent syzygies exist between the invariants of this full invariant system? These are the important questions with which we will concern ourselves subsequently.

The primary discoverer of invariant theory, Cayley, has already made the conjecture that a simultaneous system of forms has a finite full invariant system. However, he was unable to find a proof for this conjecture so he finally began to doubt its correctness again. He was only able to prove it for binary forms up to order six by assuming only one base form. Later, however, Gordan proved this theorem for an arbitrary system of simultaneous forms in two variables by using the symbolic method. But this proof is very cumbersome (compare Gordan, *Vorlesungen über Invariantentheorie*, vol. II, p. 231). Gordan's proof can also not be generalized to forms in arbitrarily many variables; he only succeeded for ternary forms of order two, three, and four (*Math. Ann.* **1**, p. 90, and the notice in Clebsch and Lindemann, *Geometrie*, vol. I, p. 274). Later,

Mertens proved finiteness in a nonsymbolic way, but also only for binary forms (cf. Crelle's *Journal für Reine und Angewandte Mathematik* **100**, p. 223.) This proof too is very cumbersome. But then, Hilbert (*Math. Ann.* **33**) gave a very simple and relatively short proof of the same fact. It is based on the representation of an invariant as a function of the differences of the roots; but for this very reason it also holds only for binary forms and cannot be generalized to forms with more variables. It was again Hilbert who then found another proof (*Math. Ann.* **36**), which can be directly generalized to arbitrary simultaneous systems of forms with arbitrarily many variables and variable sequences. This proof, which uses higher algebra, at the same time gives information about the number of syzygies and other questions that arise in finiteness considerations. In this section we first want to study the first proof of Hilbert regarding the finiteness of the full invariant system of a system of binary forms. We first prove two lemmas.

Lemma 1 *If from N arbitrary quantities $\omega, \dots,$ one forms the sum of their first, second, ..., Nth powers:*

$$\begin{gathered} \omega + \cdots, \\ \omega^2 + \cdots, \\ \cdots \\ \omega^N + \cdots, \end{gathered}$$

and if p denotes an arbitrary positive integer, then there always exists an identity of the form:

$$\omega^p = G + G^{(1)}\omega + G^{(2)}\omega^2 + \cdots + G^{(N-1)}\omega^{N-1}, \tag{1}$$

where G, $G^{(1)}$, $G^{(2)}, \dots, G^{(N-1)}$ are integral functions of the above sums of powers.

Namely, if $s_1, s_2, \dots$ are the elementary symmetric functions of the ω, then we have

$$\omega^N - s_1\omega^{N-1} + s_2\omega^{N-2} - \cdots \pm s_N = 0.$$

But $s_1, s_2, \dots, s_N$ are polynomial functions of those power sums; hence, they are of the same form as G, $G^{(1)}, \dots$. It follows that

$$\omega^N = s_1\omega^{N-1} - s_2\omega^{N-2} + \cdots \mp s_N$$

holds identically. Using ω^p, where $p \geq N$—otherwise the theorem is trivial—one can write

$$\omega^p = \omega^N \cdot \omega^{p-N} = (s_1\omega^{N-1} - s_2\omega^{N-2} + \cdots)\omega^{p-N},$$

and it is clear that by continuing in this way, one can lower the highest power of ω down to ω^{N-1}, and introduce only $\omega, \omega^2, \ldots, \omega^{N-1}$ and polynomial functions of the power sums. This proves the lemma.

Lemma 2 *A system of arbitrarily many linear homogeneous Diophantine equations has a finite number of nonnegative solutions such that every other nonnegative solution can be obtained from these linearly and homogeneously with nonnegative integer coefficients.**

The given Diophantine equations are assumed to be of the form:

$$\begin{aligned} a_1'x_1 + a_2'x_2 + \cdots + a_n'x_n &= 0, \\ a_1''x_1 + a_2''x_2 + \cdots + a_n''x_n &= 0, \\ &\cdots \\ a_1^{(m)}x_1 + a_2^{(m)}x_2 + \cdots + a_n^{(m)}x_n &= 0. \end{aligned}$$

Consider first the case of one equation. Let its positive coefficients be $\alpha_1, \alpha_2, \ldots, \alpha_\rho$; the negative ones be $-\beta_1, -\beta_2, \ldots, -\beta_\lambda$; and ξ, η replace the x. That is, the equation is

$$\alpha_1\xi_1 + \alpha_2\xi_2 + \cdots + \alpha_\rho\xi_\rho = \beta_1\eta_1 + \beta_2\eta_2 + \cdots + \beta_\lambda\eta_\lambda.$$

Suppose

$$\xi_1 = A_1,\ \xi_2 = A_2, \ldots, \xi_\rho = A_\rho, \eta_1 = B_1, \ldots, \eta_\lambda = B_\lambda$$

is an arbitrary positive solution of the equation, where all A and B are nonnegative numbers. One positive solution, however, can be given right away, namely,

$$\begin{aligned} \xi_i = \beta_k,\ \eta_k = \alpha_i;\ \xi_1 = \xi_2 = \cdots = \xi_{i-1} = \xi_{i+1} = \cdots = \xi_\rho = 0, \\ \eta_n = \eta_2 = \cdots = \eta_{k-1} = \eta_{k+1} = \cdots = \eta_\lambda = 0. \end{aligned}$$

If one takes all allowable combinations for i and k, that is, $i = 1, \ldots, \rho$; $k = 1, \ldots, \lambda$, then one obtains $\rho\lambda$ solutions, in any case a finite number. Now, let γ be the largest of all the coefficients α, β. Then all possible nonnegative solutions can be reduced, with the help of the above special solutions, to the case where either all A or all B are smaller than γ. To

* A minimal such finite set is called a *Hilbert basis* of the monoid defined by the linear Diophantine equations.

see this, suppose for instance that $A_1 > \gamma$, $B_2 > \gamma$. Then we subtract from the given solution the special solution $\xi_1 = \beta_2$, $\eta_2 = \alpha_1$; the new solution is then still nonnegative, and this process can be repeated until all A or B have become less than γ. So, we can restrict ourselves now to those solutions, say, for which all A have become less than γ. But since $\alpha < \gamma$ for all α and $\xi < \gamma$ for all relevant ξ, it follows that

$$\rho\gamma^2 > \beta_1 B_1 + \beta_2 B_2 + \cdots + \beta_\lambda B_\lambda,$$

and hence certainly $A < \gamma$ and $B < \rho\gamma^2$ for all A and B. Thus, we only need to consider those A and B. But the number of solutions for which A and B are smaller than a given finite number is certainly finite. Hence the lemma is proven for one form. An inductive step from $m-1$ to m shows that it is then valid in general. Because, if x_1', $x_2', \ldots;$ x_1'', $x_2'', \ldots; \ldots$ form the finite system of solutions for the first equation, then an arbitrary solution of the first equation is representable in the form

$$\begin{aligned} x_1 &= y_1 x_1' + y_2 x_1'' + \cdots, \\ x_2 &= y_1 x_2' + y_2 x_2'' + \cdots, \\ &\cdots, \end{aligned}$$

where the y have to be integers. If one substitutes these values in the remaining $m-1$ equations, then one obtains for y the exact same problem as before for the x. In this way, one can reduce the proof of the lemma for arbitrarily many equations to the case of one equation; but this case we have dealt with. The lemma is therefore valid in general.

Both lemmas, which we have proven, are similar to the invariant theorem in question in the sense that both assert the finiteness of certain closed systems.

Lecture XXXIII (July 6, 1897)

Let us now proceed to the proof of the theorem. We prove it for a base form

$$f = a_0(x_1 - \alpha_1 x_2)(x_1 - \alpha_2 x_2) \cdots (x_1 - \alpha_n x_2).$$

We abbreviate

$$\alpha_i - \alpha_k = (i, k).$$

Each invariant of f can be represented in the form (Section I.11):

$$\mathcal{I} = (1,2)^{e^{1,2}}(1,3)^{e^{1,3}}(2,3)^{e^{2,3}} \cdots (n-1,n)^{e^{n-1,n}} + \cdots. \qquad (2)$$

One is supposed to form sufficiently many terms of this form so that the function becomes symmetric in the α. And the $e^{1,2}$, $e^{1,3}, \ldots$, denote nonnegative integers. But every α has to appear the same number of times in this representation (Section I.11); hence, if we set $e^{i,k} = e^{k,i}$, the following equations must hold:

$$\begin{aligned} e^{1,2} + e^{1,3} + \cdots + e^{1,n} &= e^{2,1} + e^{2,3} + \cdots + e^{2,n} \\ &\cdots \\ &= e^{n,1} + e^{n,2} + \cdots + e^{n,n-1}. \end{aligned} \tag{3}$$

All positive integers which satisfy these $n-1$ homogeneous Diophantine equations, and only those, give invariants of the form f. The number of unknowns is $n(n-1)/2$. These equations have a finite system of positive integer solutions, through which all others can be expressed as linear combinations with positive integer coefficients. Let this finite system of solutions be

$$\begin{gathered} e_1^{1,2},\ e_1^{1,3},\ \ldots,\ e_1^{2,3},\ \ldots,\ e_1^{n-1,n}; \\ e_2^{1,2},\ e_2^{1,3},\ldots,\ e_2^{2,3},\ \ldots,\ e_2^{n-1,n}; \\ \cdots \\ e_m^{1,2},\ e_m^{1,3},\ \ldots,\ e_m^{2,3},\ \ldots,\ e_m^{n-1,n}. \end{gathered}$$

Every other solution can then be constructed from these with positive integer coefficients; for instance, let

$$\begin{aligned} e^{1,2} &= p_1 e_1^{1,2} + p_2 e_2^{1,2} + \cdots + p_m e_m^{1,2}, \\ e^{1,3} &= p_1 e_1^{1,3} + p_2 e_2^{1,3} + \cdots + p_m e_m^{1,3}, \\ &\cdots \\ e^{n-1,n} &= p_1 e_1^{n-1,n} + p_2 e_2^{n-1,n} + \cdots + p_m e_m^{n-1,n}. \end{aligned} \tag{4}$$

We now form the following quantities:

$$\begin{aligned} \omega_1 &= (1,2)^{e_1^{1,2}} (1,3)^{e_1^{1,3}} \cdots (n-1,n)^{e_1^{n-1,n}}, \\ &\cdots \\ \omega_m &= (1,2)^{e_m^{1,2}} (1,3)^{e_m^{1,3}} \cdots (n-1,n)^{e_m^{n-1,n}}. \end{aligned} \tag{5}$$

Then, evidently

$$\mathcal{I}_{\pi_1 \cdots \pi_m} = \omega_1^{\pi_1} \omega_2^{\pi_2} \cdots \omega_m^{\pi_m} + \cdots \tag{6}$$

is an invariant, where all subsequent terms are derived from the one that is written out by permuting the roots in all possible ways. We form all those invariants for which $\pi \leq n! = N$. This is in any case a finite number.

Each of the m products (5) gives rise to $N-1$ further expressions through permutation of the numbers $1, 2, 3, \ldots, n$ in the parentheses. Then, according to the first lemma, we have equations of the following kind:

$$\begin{aligned} \omega_1^{p_1} &= G_1 + G_1^{(1)}\omega_1 + G_1^{(2)}\omega_1^2 + \cdots + G_1^{(N-1)}\omega_1^{N-1}, \\ &\cdots \\ \omega_m^{p_m} &= G_m + G_m^{(1)}\omega_m + G_m^{(2)}\omega_m^2 + \cdots + G_m^{(N-1)}\omega_m^{N-1}, \end{aligned} \tag{7}$$

where the $G_1, G_2, \ldots, G_m$, respectively, are polynomial functions of the power sums:

$$\begin{gathered} \omega_1 + \cdots, \quad \omega_1^2 + \cdots, \quad \omega_1^N + \cdots; \\ \cdots \\ \omega_m + \cdots, \quad \omega_m^2 + \cdots, \quad \omega_m^N + \cdots. \end{gathered} \tag{8}$$

Note that the power sums (8) are invariants belonging to system (6).

But now we directly obtain the most general expression for an invariant from (2), (4), (5) to be

$$\mathcal{I} = \omega_1^{p_1}\omega_2^{p_2}\cdots\omega_m^{p_m} + \cdots.$$

If we substitute herein the values (7), then we realize immediately that $\mathcal{I}$ has to be a polynomial function of the power sums (8) and the quantities $\omega_1, \ldots, \omega_1^{N-1}, \ldots$. But since $\mathcal{I}$ is symmetric in the roots and since the exponents never get larger than N, it follows that every invariant has to be a polynomial function of the invariants (6). This completes the proof of the finiteness of the full invariant system.

It hardly needs mentioning that, while the system (6) is a full invariant system, it is not the smallest; the number of mutually independent invariants will be much smaller. But we only aimed at proving finiteness in principle.

The proof can be generalized without effort to a simultaneous system of binary forms. One only has to replace (2) by a symmetric integral function whose leading term contains all root differences of the base forms (including differences of roots of different forms) in such a way that the roots of those linear factors belonging to the same base form occur with the same degree. This characteristic property of a simultaneous invariant finds its expression in a system of linear Diophantine equations that plays the role of the system of equations (3). Aside from that, the same conclusions hold.

II.2 A generalizable proof for the finiteness of the full invariant system

We now turn to Hilbert's second proof of the finiteness of the full invariant system. This proof is very important because, as already mentioned, it can be generalized to forms with arbitrarily many variables. Here too we will consider binary forms only, but one can easily convince oneself that every step we make can be made completely analogously for forms with more than two variables.

For the moment we need to concern ourselves with a new operational symbol, which we define as follows. Let

$$f(x_1, x_2) = a_0 x_1^n + \binom{n}{1} a_1 x_1^{n-1} x_2 + \cdots + a_n x_2^n$$

be the given base form. We transform it via the transformation

$$\begin{aligned} x_1 &= \alpha y_1 + \beta y_2, \\ x_2 &= \gamma y_1 + \delta y_2 \end{aligned}$$

into

$$g(y_1, y_2) = b_0 y_1^n + \binom{n}{1} b_1 y_1^{n-1} y_2 + \cdots + b_n y_2^n.$$

The b contain α, β, γ, δ with degree n. Thus, we can differentiate them with respect to these variables. We therefore want to introduce the following notation:

$$\Omega = \frac{\partial^2}{\partial\alpha\partial\delta} - \frac{\partial^2}{\partial\beta\partial\gamma}.$$

The operation Ω proves to be of particular importance, especially in light of the following.

Lecture XXXIV (July 8, 1897)

Theorem *Let $\mathcal{F}(b)$ be a polynomial function of the transformed coefficients b. If one applies the operation Ω p times to the expression $\epsilon^\pi \mathcal{F}(b)$, considered as a function of a, α, β, γ, δ, where*

$$\epsilon = \alpha\delta - \beta\gamma$$

is the substitution determinant and π *is an integer, until* $\Omega^p(\epsilon^\pi \mathcal{F}(b))$ *does not contain the* α, β, γ, δ *anymore, then one obtains an invariant:*

$$\Omega^p(\epsilon^\pi \cdot \mathcal{F}(b)) = \mathcal{I}(a)$$

of weight $p - \pi$.

We prove this theorem as follows. We consider another linear transformation, independent of the first one:

$$y_1 = \alpha' z_1 + \beta' z_2,$$
$$y_2 = \gamma' z_1 + \delta' z_2.$$

Thereby, $f(y_1, y_2) = a_0 y_1^n + \cdots$ is transformed into

$$f(y_1, y_2) = g'(z_1, z_2) = b_0' z_1^n + \cdots.$$

Suppose furthermore that

$$\Omega' = \frac{\partial^2}{\partial\alpha'\partial\delta'} - \frac{\partial^2}{\partial\beta'\partial\gamma'},$$
$$\epsilon' = \alpha'\delta' - \beta'\gamma'.$$

Moreover, we consider a third linear transformation that is composed of the previous two, as discussed in Section I.2.

Let this transformation be

$$x_1 = \alpha'' z_1 + \beta'' z_2,$$
$$x_2 = \gamma'' z_1 + \delta'' z_2,$$

where we set

$$\alpha'' = \alpha\alpha' + \beta\gamma',$$
$$\beta'' = \alpha\beta' + \beta\delta',$$
$$\gamma'' = \gamma\alpha' + \delta\gamma',$$
$$\delta'' = \gamma\beta' + \delta\delta',$$

so that

$$\epsilon'' = \alpha''\delta'' - \beta''\gamma'' = \epsilon \cdot \epsilon',$$

and set

$$\Omega'' = \frac{\partial^2}{\partial\alpha''\partial\delta''} - \frac{\partial^2}{\partial\beta''\partial\gamma''}.$$

Under this transformation $f(x_1, x_2)$ is transformed into

$$f(x_1, x_2) = g''(z_1, z_2) = b_0'' z_1^n + \cdots.$$

The expression

$$\mathcal{I}(a) = \Omega^p\big(e^\pi \mathcal{F}(b)\big)$$

depends now only on the a, but not anymore on the transformation coefficients α, β, γ, δ; thus, it is irrelevant which transformation we use to form it. Hence we also have

$$\mathcal{I}(a) = \Omega'^p\big(\epsilon'^\pi \mathcal{F}(b')\big).$$

But the b' depend on the a in exactly the same way as the b'' depend on the b. Therefore,

$$\mathcal{I}(b) = \Omega'^p\big(\epsilon'^\pi \mathcal{F}(b'')\big).$$

From this it follows in turn that

$$\epsilon^\pi \mathcal{I}(b) = \Omega'^p\big(\epsilon^\pi \epsilon'^\pi \mathcal{F}(b'')\big),$$

since ϵ^π is a constant with respect to the operation Ω'. Therefore, one has

$$\epsilon^\pi \mathcal{I}(b) = \Omega'^p\big(\epsilon''^\pi \mathcal{F}(b'')\big).$$

To rewrite this identity further we use another property of the operation Ω; namely, the following:

Theorem *If B is a function of the α'', β'', γ'', δ'' that does not contain the α', β', γ', δ' outside of these expressions, then*

$$\Omega'(B) = \epsilon \cdot \Omega''(B).$$

The proof of this theorem is a direct computation. Namely, we have

$$\Omega'(B) = \frac{\partial^2 B}{\partial\alpha'\partial\delta'} - \frac{\partial^2 B}{\partial\beta'\partial\gamma'} = \frac{\partial}{\partial\delta'}\left(\frac{\partial B}{\partial\alpha'}\right) - \frac{\partial}{\partial\gamma'}\left(\frac{\partial B}{\partial\beta'}\right).$$

Now we have

$$\frac{\partial B}{\partial\alpha'} = \alpha\,\frac{\partial B}{\partial\alpha''} + \gamma\,\frac{\partial B}{\partial\gamma''},$$

$$\frac{\partial B}{\partial\beta'} = \alpha\,\frac{\partial B}{\partial\beta''} + \gamma\,\frac{\partial B}{\partial\delta''},$$

whence

$$\frac{\partial^2 B}{\partial\alpha'\partial\delta'} = \alpha\left\{\beta\,\frac{\partial^2 B}{\partial\alpha''\partial\beta''} + \delta\,\frac{\partial^2 B}{\partial\alpha''\partial\delta''}\right\}$$

$$+\gamma\left\{\beta\,\frac{\partial B}{\partial\beta''\partial\gamma''} + \delta\,\frac{\partial B}{\partial\gamma''\partial\delta''}\right\},$$

$$-\frac{\partial^2 B}{\partial\beta'\partial\gamma'} = -\alpha\left\{\beta\,\frac{\partial^2 B}{\partial\alpha''\partial\beta''} + \delta\,\frac{\partial^2 B}{\partial\beta''\partial\gamma''}\right\}$$

$$-\gamma\left\{\beta\,\frac{\partial^2 B}{\partial\alpha''\partial\delta''} + \delta\,\frac{\partial^2 B}{\partial\gamma''\partial\delta''}\right\}.$$

And from this it follows indeed that

$$\frac{\partial^2 B}{\partial\alpha'\partial\delta'} - \frac{\partial^2 B}{\partial\beta'\partial\gamma'} = (\alpha\delta - \beta\gamma)\left(\frac{\partial^2 B}{\partial\alpha''\partial\delta''} - \frac{\partial^2 B}{\partial\beta''\partial\gamma''}\right),$$

that is,

$$\Omega'(B) = \epsilon\cdot\Omega''(B).$$

Repeated application of this theorem gives:

$$\Omega'^p(B) = \epsilon^p\cdot\Omega''^p(B),$$

where B is a function as described above. One can even consider B to be arbitrary here; one only has to make sure that on the left-hand side one considers B as a function of the variables $\alpha,\ldots;\ \alpha',\ldots$, which are to be considered independent. On the right-hand side, however, B is to be considered as a function of the variables $\alpha,\ldots;\ \alpha'',\ldots$, which are now to be considered independent.

According to this theorem, the formula

$$\epsilon^\pi\mathcal{I}(b) = \Omega'^p\big(\epsilon'^\pi\mathcal{F}(b'')\big)$$

now becomes

$$\epsilon^\pi\mathcal{I}(b) = \epsilon^p\Omega''^p\big(\epsilon''^\pi\mathcal{F}(b'')\big),$$

and since, according to an earlier remark, the right-hand side is equal to $\epsilon^p\mathcal{I}(a)$, we have

$$\mathcal{I}(b) = \epsilon^{p-\pi}\mathcal{I}(a),$$

as asserted.

One can also prove the converse.

Theorem *For each invariant $\mathcal{I}$ one can find a function $\mathcal{F}$ such that, for suitable π and p, the equation*

$$\mathcal{I}(a) = \Omega^p\big(\epsilon^\pi \mathcal{F}(b)\big)$$

holds.

One only needs to take

$$\mathcal{F} = C \cdot \mathcal{I},$$

where C is a constant to be determined. Since then

$$\mathcal{I}(b) = \epsilon^p \mathcal{I}(a),$$

it follows from

$$\Omega^{p_1}\big(C\epsilon^p \mathcal{I}(a)\big) = \mathcal{I}(a)$$

that

$$\Omega^{p_1}(C\epsilon^p) = 1,$$

or, since this implies $p = p_1$, that

$$C = \frac{1}{\Omega^p \epsilon^p}.$$

But this is always possible since $\Omega^p \epsilon^p$ *is always different from zero.* This last statement is a very essential fact that we will use often. To prove it we remark that Ω^2 can be written in the form

$$\Omega^2 = \left(\frac{\partial^2}{\partial\alpha\partial\delta}\right)^2 - 2\,\frac{\partial^2}{\partial\alpha\partial\delta}\,\frac{\partial^2}{\partial\beta\partial\gamma} + \left(\frac{\partial^2}{\partial\beta\partial\gamma}\right)^2,$$

an expression which is to be considered symbolic, however, that is, such that one adds the superscripts in the numerator in such a way that, for instance,

$$\left(\frac{\partial^2}{\partial\alpha\partial\delta}\right)^2 = \frac{\partial^4}{\partial\alpha^2\partial\delta^2}, \qquad \frac{\partial^2}{\partial\alpha\partial\delta}\,\frac{\partial^2}{\partial\beta\partial\gamma} = \frac{\partial^4}{\partial\alpha\partial\beta\partial\gamma\partial\delta}.$$

An inductive step from p to $p+1$ shows easily that, using analogous notation, we obtain for higher differential quotients that

$$\begin{aligned}\Omega^p = {} & \left(\frac{\partial^2}{\partial\alpha\partial\delta}\right)^p - \binom{p}{1}\left(\frac{\partial^2}{\partial\alpha\partial\delta}\right)^{p-1}\left(\frac{\partial^2}{\partial\beta\partial\gamma}\right)^1 \\ & + \binom{p}{2}\left(\frac{\partial^2}{\partial\alpha\partial\delta}\right)^{p-2}\left(\frac{\partial^2}{\partial\beta\partial\gamma}\right)^2 - \cdots + (-1)^p\left(\frac{\partial^2}{\partial\beta\partial\gamma}\right)^p.\end{aligned}$$

On the other hand we have

$$\epsilon^p = (\alpha\delta)^p - \binom{p}{1}(\alpha\delta)^{p-1}\beta\gamma + \binom{p}{2}(\alpha\delta)^{p-2}(\beta\gamma)^2 - \cdots + (-1)^p(\beta\gamma)^p.$$

Applying Ω^p to this expression, only those terms contribute something that correspond to a summand in Ω^p. The others don't contain either α, δ or β, γ to a sufficiently high power, and therefore give zero. Thus, we obtain for the desired $\Omega^p(\epsilon^p)$:

$$\Omega^p\epsilon^p = p!^2 + \binom{p}{1}^2 (p-1)!^2 1!^2 + \binom{p}{2}^2 (p-2)!^2 2!^2 + \cdots + p!^2,$$

or

$$\Omega^p\epsilon^p = (p+1)p!^2.$$

This proves our assertion.

Lecture XXXV (July 9, 1897)

For the proof of the finiteness of the full invariant system we now have to derive another theorem that represents a general fact about finiteness.*

General Finiteness Theorem *Let $\mathcal{F}_1, \mathcal{F}_2, \mathcal{F}_3, \ldots$ be an infinite sequence of forms in the n variables $x_1, x_2, \ldots x_n$. Then there always exists an integer m such that each form in the sequence can be expressed as*

$$\mathcal{F} = \mathcal{A}_1\mathcal{F}_1 + \mathcal{A}_2\mathcal{F}_2 + \cdots + \mathcal{A}_m\mathcal{F}_m,$$

where $\mathcal{A}_1, \mathcal{A}_2, \ldots, \mathcal{A}_m$ are suitable forms of the same n variables.

Prof. Hilbert has discussed the importance and impact of his theorem in *Math. Ann.* **36**. Indeed, it has important applications for many purposes, and we will see that, in particular, we can use it together with the other observations of this section, to prove the finiteness of the full invariant system, in a way that is easily seen to be generalizable. We must now prove the theorem.

Note also that the statement of the theorem assumes that the given sequence of forms $\mathcal{F}_1, \mathcal{F}_2, \mathcal{F}_3, \ldots$ is a countable set, that is, one can think of it as ordered in some way, according to some given rule, and that it is given in that order. But there are no additional hypotheses.

In the simplest case $n = 1$, the theorem is clear. Each $\mathcal{F}$ has the form

* The following theorem is now known as the Hilbert Basis Theorem.

cx^r, where c is a constant. Let $c_1x^{r_1}$ be the first form of the sequence with a coefficient different from zero. We then look for the next form in the sequence whose order is less than r_1; if there is no such form, we retain $c_1x^{r_1}$. But if there is one, say $c_2x^{r_2}$, then we proceed to the next form in the sequence whose order is less than r_2. If we continue in this manner, then we finally arrive at a form $c_ix^{r_i} = \mathcal{F}_m$ in the sequence with the property that none of the subsequent forms have order less than r_i. Every form is then divisible by $\mathcal{F}_m$, and so m is an integer which satisfies the assertion of the theorem.

For $n = 2$, we can argue analogously. We sketch the proof for binary forms as follows. Suppose that the first two forms have a common linear factor, and assume further that this factor is contained in every subsequent form of the sequence. Arguing as in the previous case, there is a form in which this factor appears to the lowest power. We divide all forms by it and obtain a new sequence that does not have this common factor anymore. But it can have another factor in common, which we also remove in the same way. But since the forms have finite order, there can only be a finite number of common factors. Hence, after a finite number of operations, we reach a sequence of forms such that no factor is common to all forms. Let this be

$$\mathcal{F}'_1, \mathcal{F}'_2, \mathcal{F}'_3, \dots .$$

We proceed to a point in the sequence such that the first part

$$\mathcal{F}'_1, \mathcal{F}'_2, \dots, \mathcal{F}'_m$$

has the property that the forms have no common factor. But then we can construct two forms Φ_1, Φ_2 so that

$$\Phi_1 = A_1\mathcal{F}'_1 + A_2\mathcal{F}'_2 + \cdots + A_m\mathcal{F}'_m,$$
$$\Phi_2 = B_1\mathcal{F}'_1 + B_2\mathcal{F}'_2 + \cdots + B_m\mathcal{F}'_m$$

have no common factor, where the A and B are forms of the prescribed kind. But then, any arbitrary binary form $\mathcal{F}$, whose order is not smaller than the sum r of the orders of Φ_1 and Φ_2, can be put in the form

$$\mathcal{F} = A\Phi_1 + B\Phi_2,$$

where A and B are suitable forms. As far as the forms of order smaller than r are concerned, one can in any case find a finite system with the property that all others can be expressed by it. Finally, to recover the original form $\mathcal{F}$ we only need to multiply the obtained identities with the common factor.

For $n = 3$, the proof of our theorem would already encounter sizeable difficulties, which would only increase as soon as we further increase the number of variables. We will therefore take a different route to the proof of the theorem, namely, an inductive argument from $n-1$ variables to n variables.

Let $\mathcal{F}_1, \mathcal{F}_2, \mathcal{F}_3, \ldots$ be the given sequence of forms in n variables, and let $\mathcal{F}_1$ be a nonvanishing form of order r. Then we first determine a linear substitution of the variables $x_1, x_2, \ldots, x_n$ which has a nonzero determinant and transforms $\mathcal{F}_1$ into a form $\mathcal{G}_1$ of the variables $y_1, y_2, \ldots, y_n$ in such a way that the coefficient of y_n^r in the form $\mathcal{G}_1$ is nonzero. Suppose this linear substitution transforms $\mathcal{F}_2, \mathcal{F}_3, \ldots$, respectively, into forms $\mathcal{G}_2, \mathcal{G}_3, \ldots$. If we now consider a relation of the form

$$\mathcal{G}_s = B_1\mathcal{G}_1 + B_2\mathcal{G}_2 + \cdots + B_m\mathcal{G}_m,$$

where s is some subscript and $B_1, B_2, \ldots, B_m$ are forms in the variables $y_1, y_2, \ldots, y_n$, then it is transformed by the inverse linear substitution into a relation of the form

$$\mathcal{F}_s = A_1\mathcal{F}_1 + A_2\mathcal{F}_2 + \cdots + A_m\mathcal{F}_m,$$

where $A_1, A_2, \ldots, A_m$ are forms in the original variables $x_1, x_2, \ldots, x_n$. Hence, the theorem is proven for the forms $\mathcal{F}$, if it is proven for the forms $\mathcal{G}$.

Since the coefficient of y_n^r in $\mathcal{G}_1$ is different from zero, one can reduce the degree of any form $\mathcal{G}_s$ in the sequence $\mathcal{G}$ with respect to the variable y_n below r by multiplying $\mathcal{G}_1$ with a suitable form B_s and subtracting the product from $\mathcal{G}_s$; this can be seen by direct division. We therefore set, for an arbitrary s:

$$\mathcal{G}_s = B_s\mathcal{G}_1 + g_{s1}y_n^{r-1} + g_{s2}y_n^{r-2} + \cdots + g_{sr},$$

where B_s is a form in the n variables $y_1, \ldots, y_n$, while the forms $g_{s1}, g_{s2}, \ldots, g_{sr}$ contain only the $n-1$ variables $y_1, \ldots, y_{n-1}$.

We now assume that our theorem is already proven for forms in $n-1$ variables, and apply it to the sequence of forms $g_{11}, g_{21}, g_{31}, \ldots$. Hence, one can determine a number μ such that for every value of s we can set

$$g_{s1} = b_{s1}g_{11} + b_{s2}g_{21} + \cdots + b_{s\mu}g_{\mu 1} = l_s(g_{11}, g_{21}, \ldots, g_{\mu 1}),$$

where $b_{s1}, \ldots, b_{s\mu}$ are forms of the $n-1$ variables $y_1, \ldots, y_{n-1}$. We now construct the forms

$$g_{st}^{(1)} = g_{st} - l_s(g_{1t}, \ldots, g_{\mu t}) \qquad (t = 1, 2, \ldots, r), \tag{1}$$

from which we obtain in particular, for $t = 1$, that

$$g_{s1}^{(1)} = 0.$$

We now again use our theorem for $n - 1$ variables and apply it to the form sequence $g_{12}^{(1)}, g_{22}^{(1)}, g_{32}^{(1)}, \ldots$. One can then determine an integer $\mu^{(1)}$ such that, for every value of s, we have a relation of the form

$$\begin{aligned} g_{s2}^{(1)} &= b_{s1}^{(1)} g_{12}^{(1)} + b_{s2}^{(1)} g_{22}^{(1)} + \cdots + b_{s\mu^{(1)}}{}^{(1)} g_{\mu^{(1)}2}{}^{(1)} \\ &= l_s^{(1)} \left(g_{12}^{(1)}, g_{22}^{(1)}, \ldots, g_{\mu^{(1)}2}{}^{(1)} \right), \end{aligned}$$

where again $b_{s1}^{(1)}, \ldots, b_{s\mu^{(1)}}{}^{(1)}$ are forms in the $n - 1$ variables $y_1, \ldots, y_{n-1}$. We set

$$g_{st}^{(2)} = g_{st}^{(1)} - l_s^{(1)} \left(g_{1t}^{(1)}, g_{2t}^{(1)}, \ldots, g_{\mu^{(1)}t}{}^{(1)} \right) \qquad (t = 1, 2, \ldots, r), \tag{2}$$

from which it follows that, for $t = 1, 2$, we have

$$g_{s1}^{(2)} = 0, \qquad g_{s2}^{(2)} = 0.$$

The application of our theorem to the form sequence $g_{13}^{(2)}, g_{23}^{(2)}, g_{33}^{(2)}, \ldots$ leads to the relation

$$g_{s3}^{(2)} = l_s^{(2)} \left(g_{13}^{(2)}, g_{23}^{(2)}, \ldots, g_{\mu^{(2)}3}{}^{(2)} \right).$$

If we then set

$$g_{st}^{(3)} = g_{st}^{(2)} - l_s^{(2)} \left(g_{1t}^{(2)}, g_{2t}^{(2)}, \ldots, g_{\mu^{(2)}t}{}^{(2)} \right) \qquad (t = 1, 2, \ldots, r) \tag{3}$$

then it follows, in particular, that

$$g_{s1}^{(3)} = 0, \qquad g_{s2}^{(3)} = 0, \qquad g_{s3}^{(3)} = 0,$$

and, continuing in this way, one ultimately obtains

$$g_{st}^{(r-1)} = g_{st}^{(r-2)} - l_s^{(r-2)} \left(g_{1t}^{(r-2)}, g_{2t}^{(r-2)}, \ldots, g_{\mu^{(r-2)}t}^{(r-2)} \right) \qquad (t = 1, 2, \ldots, r), \tag{4}$$

$$g_{s1}^{(r-1)} = 0, \; g_{s2}^{(r-1)} = 0, \ldots, g_{s,r-1}^{(r-1)} = 0.$$

And finally:

$$g_{sr}^{(r-1)} = l_s^{(r-1)} \left(g_{1r}^{(r-1)}, g_{2r}^{(r-1)}, \ldots, g_{\mu^{(r-1)}r}^{(r-1)} \right)$$

from which

$$0 = g_{st}^{(r-1)} - l_s^{(r-1)} \left(g_{1t}^{(r-1)}, g_{2t}^{(r-1)}, \ldots, g_{\mu^{(r-1)}t}^{(r-1)} \right) \quad (t = 1, 2, \ldots, r) \tag{5}$$

follows. Addition of the equations (1), (2), (3), (4), (5) results in

$$g_{st} = l_s\left(g_{1t}, g_{2t}, \ldots, g_{\mu t}\right) + l_s^{(1)}\left(g_{1t}^{(1)}, g_{2t}^{(1)}, \ldots, g_{\mu^{(1)}t}^{(1)}\right) + \cdots$$
$$+ l_s^{(r-1)}\left(g_{1t}^{(r-1)}, g_{2t}^{(r-1)}, \ldots g_{\mu^{(r-1)}t}^{(r-1)}\right) \qquad (t = 1, 2, \ldots, r).$$

On the right-hand side we can replace the forms

$$g_{1t}^{(r-1)}, g_{2t}^{(r-1)}, \ldots, g_{\mu^{(r-1)}t}^{(r-1)}, \ldots, g_{1t}^{(1)}, g_{2t}^{(1)}, \ldots, g_{\mu^{(1)}t}^{(1)}$$

by linear combinations of the forms $g_{1t}, g_{2t}, \ldots, g_{mt}$ through repeated application of the equations (1)–(4), where m denotes the largest of the integers $\mu, \mu^{(1)}, \ldots, \mu^{(r-1)}$. In this way we obtain from the last formula a system of equations of the form:

$$g_{st} = c_{s1}g_{1t} + c_{2s}g_{2t} + \cdots + c_{sm}g_{mt}$$
$$= k_s\left(g_{1t}, g_{2t}, \ldots, g_{mt}\right) \qquad (t = 1, 2, \ldots, r),$$

where $c_{s1}, c_{s2}, \ldots, c_{sm}$ are again forms in the $n-1$ variables $y_1, y_2, \ldots,$ y_{n-1}. If we multiply the last formula by y_n^{r-t} and add the resulting equations for $t = 1, 2, \ldots, r$, then we obtain the equation

$$\mathcal{G}_s - B_s\mathcal{G}_1 = k_s\left(\mathcal{G}_1 - B_1\mathcal{G}_1, \mathcal{G}_2 - B_2\mathcal{G}_1, \ldots, \mathcal{G}_m - B_m\mathcal{G}_1\right) ,$$

since

$$g_{s1}y_n^{r-1} + g_{s2}y_n^{r-2} + \cdots + g_{sr} = \mathcal{G}_s - B_s\mathcal{G}_1 .$$

If C_s denotes a suitable form in the n variables $y_1, y_2, \ldots, y_n$, then this equation becomes

$$\mathcal{G}_s = C_s\mathcal{G}_1 + k_s\left(\mathcal{G}_1, \mathcal{G}_2, \ldots, \mathcal{G}_m\right) = \mathcal{L}_s\left(\mathcal{G}_1, \mathcal{G}_2, \ldots, \mathcal{G}_m\right);$$

that is, the integer m is a number satisfying the requirements of the theorem for the form sequence $\mathcal{G}_1, \mathcal{G}_2, \mathcal{G}_3, \ldots,$ and hence also for the sequence $\mathcal{F}_1, \mathcal{F}_2, \mathcal{F}_3, \ldots$. Therefore, the theorem is valid for n variables if it holds for $n-1$ variables; and since it is proven for one variable, in holds in general.

Lecture XXXVI (July 12, 1897)

With the help of the previous two theorems it is now easy to prove the finiteness of the full invariant system. We consider the complete system of invariants of the binary form of order n

$$f(x_1, x_2) = a_0x_1^n + \cdots + a_nx_2^n.$$

The invariants are forms in the $n+1$ variables $a_0, \ldots, a_n$. They clearly form a countable set, if we first select only the linearly independent ones. We are allowed to do this, since the additional linear functions of the given invariants do not contribute to the full invariant system. But then, the number of invariants of a given degree is finite, because the number of terms of a given degree is finite, and the worst possible case—which in fact never occurs—is that there exist as many invariants as there are terms of a given degree. Then each of these terms would be a linear function of the invariants, hence an invariant itself; and so all conceivable homogeneous functions of any degree would be linear functions of the above invariants, hence invariants themselves.

But since the number of invariants of a given degree is in any case finite—incidentally, one can see this directly from the enumeration calculus, although we want to phrase everything now so as to make its generalizability immediately apparent—one can order the invariants of a given degree in some way, for instance, as it is done with division of algebraic expressions. That is, one first takes the invariants with a_0^g (where g is the degree), then those with $a_0^{g-1}a_1, a_0^{g-1}a_2, \ldots, a_0^{g-1}a_n$, then those with $a_0^{g-2}a_1^2, a_0^{g-2}a_1a_2, \ldots$. Using this order, one can first write the invariants of degree one, then those of degree two, etc. (The constant is expressly omitted.) Suppose the invariants, in this order, are

$$i_1, i_2, i_3, \ldots .$$

Then our finiteness theorem says that an arbitrary invariant i in this sequence can be put in the form:

$$i = \mathcal{A}_1 i_1 + \mathcal{A}_2 i_2 + \cdots + \mathcal{A}_m i_m,$$

where $i_1, i_2, \ldots, i_m$ are the first m invariants, and m is a finite number, and where $\mathcal{A}_1, \mathcal{A}_2, \ldots, \mathcal{A}_m$ are forms in the a. But then we assert that i can be expressed integrally and rationally through these m invariants $i_1, i_2, \ldots, i_m$. The above equation is an identity. So, if b are the transformed coefficients, then we have in any case:

$$i(b) = \mathcal{A}_1(b) \cdot i_1(b) + \mathcal{A}_2(b) \cdot i_2(b) + \cdots + \mathcal{A}_m(b) \cdot i_m(b).$$

Now, $i, i_1, \ldots, i_m$ are invariants; let their weights be $\pi, \pi_1, \ldots, \pi_m$, respectively; then, in the notation of the beginning of the section, we have

$$\epsilon^{\pi} i(a) = \mathcal{A}_1(b) \cdot \epsilon^{\pi_1} i_1(a) + \mathcal{A}_2(b) \cdot \epsilon^{\pi_2} i_2(a) + \cdots + \mathcal{A}_m(b) \cdot \epsilon^{\pi_m} i_m(a).$$

If we apply the operation Ω to this identity π times, then it follows* that

$$i(a)\Omega^{\pi}\epsilon^{\pi} = i_1(a)\Omega^{\pi}\left(\epsilon^{\pi_1}\mathcal{A}_1(b)\right) + \cdots + i_m(a)\Omega^{\pi}\left(\epsilon^{\pi_m}\mathcal{A}_m(b)\right),$$

and therefore** (Lecture XXXIV):

$$i(a) = \mathcal{I}_1(a)\cdot i_1(a) + \mathcal{I}_2(a)\cdot i_2(a) + \cdots + \mathcal{I}_m(a)\cdot i_m(a),$$

where, according to our theorem about Ω, the expressions of the coefficients $\mathcal{I}_1(a), \mathcal{I}_2(a), \ldots, \mathcal{I}_m(a)$ are again invariants. Because it is evident that after π-fold application of Ω to $\epsilon^{\pi_1}\mathcal{A}_1(b)$ exactly the $\alpha, \beta, \gamma, \delta$ drop out, since ϵ^{π_1} contains them to the same degree as $i_1(b)$—because of $i_1(b) = \epsilon^{\pi_1} i_1(a)$—and hence $\mathcal{A}_1(b)\cdot\epsilon^{\pi_1}$ contains them to the same degree as $i_1(b)\mathcal{A}_1(b)$.

Now $i_1(a)$, $i_2(a), \ldots, i_m(a)$ have at least degree one, so $\mathcal{I}_1$, $\mathcal{I}_2, \ldots, \mathcal{I}_m$ have certainly lower degree than $i(a)$. We can do the same with $\mathcal{I}_1(a)$, $\ldots, \mathcal{I}_m(a)$ as with $i(a)$, and can clearly continue this process until the degree of the invariants $\mathcal{I}$, used as coefficients, is smaller than that of i_m; then we have shown that each invariant has to be a polynomial function of the invariants $i_1, i_2, \ldots, i_m$. It is, of course, possible that the number of invariants in the full invariant system is smaller than m; but this is irrelevant. We have in any case proven rigorously the following:

Theorem (Finiteness of the Full Invariant System) *Every binary form possesses a finite full invariant system such that each invariant of the form is a polynomial function of the invariants in the full invariant system.*

Lecture XXXVII (July 13, 1897)

One sees easily that the above proof can be generalized, since it is evident that the operation Ω can be generalized, in a suitable way, to forms with more than two variables, and that then theorems hold which are entirely analogous to those that we have proven for binary forms. We can even admit several sequences of variables; indeed, one can even subject them to various linear transformations without invalidating the observations. In this way, one can prove quite generally:

* Note that $\Omega^{\pi}\epsilon^{\pi}$ is a nonzero constant.

** That is, as a polynomial function.

Theorem *Given a system of base forms in arbitrarily many sequences of variables, which are subject to various linear transformations in a prescribed way, there exists a finite number of integral rational invariants through which every other integral rational invariant can be expressed integrally and rationally.*

This also implies the finiteness of the full system for covariants, since they are special cases of invariants.

The situation is different, however, as soon as one generalizes the invariant concept as in the investigation of F. Klein (1872), and Sophus Lie (*Theorie der Transformationsgruppen*, Leipzig 1888). So far, we had defined an invariant as an integral homogeneous function of the coefficients of the base forms that possesses the invariant property with respect to *all* linear transformations of the variables. But one can also follow the more general definition, based on a certain subgroup of the general group of linear transformations, and then ask about those integral homogeneous functions of the coefficients of the base forms that possess the invariant property only with respect to the substitutions of that subgroup. Even though these invariants obviously contain the invariants in the previous sense, it does not yet follow from our theorems about the finiteness of full invariant systems that for the invariants in the general sense one can always find a finite number through which every other invariant of the same kind can be expressed integrally and rationally. In certain cases, the method of proof of this section can also be applied to this new question (see Section II.8 as well as the quoted paper by Hilbert (1890)).

With each mathematical theorem, three things are to be distinguished. First, one needs to settle the basic question of whether the theorem is valid; one has to prove its existence, so to speak. Second, one can ask whether there is any way to determine how many operations are needed at the most to carry out the assertion of the theorem. Kronecker has particularly emphasized the question of whether one can carry it out in a finite number of steps. Third, it has to be actually carried out; this is the least interesting question. We illustrate these questions with an example. One can ask: Is there a place somewhere in the decimal expansion of $\pi = 3.14159\ldots$ at which there appear ten consecutive ones 1111111111? It is not improbable that this may be the case. If we assume one could prove this in some way, then one can ask the second question: Can one find a number N of which one knows that there are at least ten consecutive ones before the Nth decimal of $\pi = 3.14\ldots$? The number N can be much too large, as long as we can only prove the assertion for it. Third, one would then actually have to calculate the number N_1 so that the N_1th up to the (N_1+9)th decimal are all ones, and so that there are no ten consecutive ones appearing earlier; or one might have to calculate the first N_1+9 decimals of π.

We too are faced with a similar question. We have proven that a binary form of order n has a finite full invariant system. Suppose that the degree of the invariant of highest degree in the simplest full invariant system is N; then the question arises: If we are given the number n, can we calculate an upper bound for N? Our proof has settled the finiteness question only in principle; there is not the slightest indication that we can actually calculate such a number N. If one considers $n = 307$, for instance, then the proof does not indicate any number of which one would know that it is bigger than N. We will have opportunity in the next sections, however, to consider this question further. The third question would here amount to actually calculating the invariants which form the full invariant system. (We will not consider this question, which is pointless for base forms of higher order.)

II.3 The system of invariants $\mathcal{I}$; $\mathcal{I}_1, \mathcal{I}_2, \ldots, \mathcal{I}_k$ *

For the sake of concreteness, in the following we again fix a binary form. The following observations would hardly have to be modified if we were to generalize them in the sense of Section II.2.

The invariants form an integral domain with a finite basis, which constitutes the full invariant system. First we inquire about factorization in the integral domain. Here, we simply have the following:

Theorem *Each invariant can be factored uniquely into prime invariants.*

Indeed, let

$$i = i_1 i_2 \cdots i_h$$

be an invariant; the decomposition into factors that are polynomial functions of the coefficients $a_0, \ldots$ is well known to be unique. On the other hand, *each factor of an invariant is again an invariant,* because we have

$$i(b) = i_1(b) i_2(b) \cdots i_h(b),$$

* The results in the present and following sections were first published in Hilbert (1893).

and furthermore

$$\epsilon^\pi i(a) = \epsilon^\pi i_1(a) i_2(a) \cdots i_h(a) = i_1(b) i_2(b) \cdots i_h(b).$$

Since $\epsilon, i_1, i_2, \ldots, i_h$ are irreducible functions, one of the factors $i_1(b)$, $\ldots, i_h(b)$ necessarily has to become $\epsilon^\rho i_h(a)$ (where ρ is possibly zero) when we express the b in terms of the a. Hence, we have

$$\epsilon^\pi i(a) = i_1(b) i_2(b) \cdots i_{l-1}(b) i_{l+1}(b) \cdots i_h(b) \epsilon^\rho i_h(a).$$

Applying Cayley's Ω-process to both sides of the previous equation, we get

$$i(a) \Omega^\pi \epsilon^\pi = i_h(a) \Omega^\pi \left\{ \epsilon^\rho i_1(b) \cdots i_h(b) \right\},$$

so

$$i_1(a) i_2(a) \cdots i_{h-1}(a) = \frac{i(a)}{i_h(a)} = \frac{\Omega^\pi (\epsilon^\rho i_1(b) \cdots i_h(b))}{\Omega^\pi \epsilon^\pi}$$

is an invariant. But since the quotient of two invariants is again an invariant, if the denominator divides the numerator, the quotient $i(a)/\frac{i(a)}{i_h(a)} = i_h(a)$ is also an invariant. The conceivable case that one of the factors $i(b)$ turns into several factors $i(a)$ cannot happen because after continuing the process there would be nothing left for the last factor.

In the previous section we have studied, or at least proven the existence of, a system of invariants through which all other invariants can be expressed integrally and rationally, that is, as a polynomial function. But one can specialize the question and ask whether there is a system of invariants through which all others can be expressed integrally and algebraically, that is, by abandoning the "rational." * It is clear that the smallest such system cannot contain more invariants than the invariant system for an integral and rational expression, which is at the same time an integral algebraic expression. In general, however, the number of invariants which allow an integral algebraic expression of the others will be much smaller, since there will always be a number of syzygies that occur in large numbers for base forms of higher order. For instance, for a binary form of order one and one of order five there are twenty-three joint invariants; but by eliminating the eight quantities a, b from

* In the first case, we are looking for invariants which generate the invariant ring as an algebra; in the second case the invariant ring is integral over the algebra they generate.

the equations

$$\mathcal{I}_1 = f_1(a_0, \ldots, a_5;\ b_0, b_1),$$
$$\mathcal{I}_2 = f_2(a_0, \ldots, a_5;\ b_0, b_1),$$
$$\ldots$$
$$\mathcal{I}_{23} = f_{23}(a_0, \ldots, a_5;\ b_0, b_1)$$

one sees that there will exist at least fifteen syzygies, and there could even be more.

After these preliminary remarks, we prove the following:

Theorem *Given an arbitrary base form or system of base forms, one can always find certain invariants $\mathcal{I}_1, \mathcal{I}_2, \ldots, \mathcal{I}_K$ so that there is no algebraic relation between them and through which every other invariant can be expressed integrally and algebraically.**

That is, if i is an arbitrary invariant, then it satisfies the equation

$$i^m + \mathcal{G}_1 i^{m-1} + \mathcal{G}_2 i^{m-2} + \cdots + \mathcal{G}_m = 0,$$

where $\mathcal{G}_1, \mathcal{G}_2, \ldots, \mathcal{G}_m$ are integral rational functions of the invariants $\mathcal{I}_1, \mathcal{I}_2, \ldots \mathcal{I}_K$.

To prove the theorem, let $i_1, i_2, \ldots, i_m$ be the invariants of the full invariant system; if there does not exist a relation between them, the theorem already holds. If there exists one, however, let it be

$$R(i_1, i_2, \ldots, i_m) = 0.$$

Suppose that i_1 is of degree ν_1 in the coefficients a, i_2 of degree $\nu_2, \ldots$, i_m of degree ν_m. Since $i_1, i_2, \ldots, i_m$ are homogeneous functions, we may assume that after introducing the a, R is a homogeneous function in them, because otherwise the homogeneous parts would have to become zero individually. If we now set

$$i'_1 = i_1^{\nu/\nu_1}, i'_2 = i_2^{\nu/\nu_2}, \ldots, i'_m = i_m^{\nu/\nu_m},$$

where $\nu = \nu_1 \cdot \nu_2 \cdots \nu_m$, then all i' are homogeneous in the a of degree ν, and it is clear that the relation $R = 0$ can be brought into the form

$$R'(i'_1, i'_2, \ldots, i'_m) = 0,$$

where R' is an integral rational homogeneous function of $i'_1, i'_2, \ldots, i'_m$, of some degree e. Possibly, we have

$${i'_m}^e + \cdots = 0.$$

* This theorem amounts to applying Noether Normalization to the invariant ring.

But if the coefficient of ${i'_m}^e$ is zero, then one can find, as explained in the previous section (Lecture XXXV), a linear substitution with nonvanishing determinant such that, if

$$i'_1 = \alpha_{11} i''_1 + \cdots + \alpha_{1m} i''_m,$$
$$\cdots$$
$$i'_m = \alpha_{m1} i''_1 + \cdots + \alpha_{mm} i''_m,$$

then the coefficient of ${i''_m}^e$ in R' is equal to one. But then there exists an equation between the invariants $i''_1, \dots, i''_m$. The fact that these are invariants follows from the invertibility of the transformation, together with the fact that the invariants $i'_1, \dots, i'_m$ all have the same degree, and this equation shows that i''_m, hence i'_m and, therefore, also i_m, is an integral algebraic function of the $m-1$ invariants $i''_1, \dots, i''_{m-1}$. If there are no more syzygies between them, the theorem is proven; if there is one, then one applies the same procedure again. It is clear that by continuing this process one gives a direct proof of the theorem. It is worth remarking, incidentally, that the system of invariants $\mathcal{I}_1, \mathcal{I}_2, \dots, \mathcal{I}_K$ need not be determined uniquely; we have only proven that there is at least one such system. Furthermore, our proof shows that one can determine $\mathcal{I}_1, \mathcal{I}_2, \dots, \mathcal{I}_K$ so that they all have the same degree.

Lecture XXXVIII (July 15, 1897)

We continue with the following:

Theorem *It is always possible to add an invariant $\mathcal{I}$ to the invariants $\mathcal{I}_1, \mathcal{I}_2, \dots, \mathcal{I}_K$ such that every other invariant of the base form can be expressed rationally in terms of the invariants $\mathcal{I}; \mathcal{I}_1, \mathcal{I}_2, \dots, \mathcal{I}_K$.**

We are now dropping the "integral" from the rational expression while we omitted the "rational" earlier. We prove this theorem as follows. Let i_1, i_2 be any two invariants of the full invariant system $i_1, i_2, \dots, i_m$ through which all others can be expressed integrally and rationally. According to a theorem about algebraic functions, one can determine two constants c_1, c_2 such that i_1 and i_2 can be expressed rationally in terms of $\mathcal{I}_1, \mathcal{I}_2, \dots, \mathcal{I}_K$ and

$$i_{12} = c_1 i_1^{\alpha_1} i_2^{\alpha_2} + c_2 i_1^{\beta_1} i_2^{\beta_2} \mathcal{I}_1^{\gamma}.$$

Here $i_1^{\alpha_1} i_2^{\alpha_2}$ and $i_1^{\beta_1} i_2^{\beta_2} \mathcal{I}_1^{\gamma}$ have the same degree in the coefficients of the

* This is a special case of the Primitive Element Theorem.

base form; that is, i_{12} is an invariant (for the construction of $\alpha_1, \alpha_2, \beta_1, \beta_2$ see the treatise Hilbert (1893). Therefore, every invariant can be expressed rationally in terms of $i_{12}, \mathcal{I}_1, \mathcal{I}_2, \ldots, \mathcal{I}_K, i_3, i_4, \ldots, i_m$. This line of reasoning can be continued by first combining i_{12} and i_3 to get an invariant $i_{123}\ldots$, etc., until one finally obtains an invariant $\mathcal{I} = i_{123\cdots m}$ such that all other invariants can be expressed rationally in terms of $\mathcal{I}; \mathcal{I}_1, \mathcal{I}_2, \ldots, \mathcal{I}_K$, as asserted by the theorem. Of course, $\mathcal{I}$ satisfies an equation of the form:

$$\mathcal{I}^k + \mathcal{G}_1(\mathcal{I}_1, \ldots, \mathcal{I}_K)\mathcal{I}^{k-1} + \mathcal{G}_2(\mathcal{I}_1, \ldots)\mathcal{I}^{k-2} + \cdots + \mathcal{G}_k = 0.$$

If we assume that k is as small as possible, so that the equation above will be irreducible, then $\mathcal{I}$ is an integral algebraic function of degree k in $\mathcal{I}_1, \mathcal{I}_2, \ldots, \mathcal{I}_K$.

The invariants $\mathcal{I}; \mathcal{I}_1, \mathcal{I}_2, \ldots, \mathcal{I}_K$ determine a function field that we shall simply call the *invariant field* of the base form. The invariant field is of degree k. It is very important to study the invariant field in more detail.

As we have just shown, each invariant is an integral algebraic function of $\mathcal{I}; \mathcal{I}_1, \mathcal{I}_2, \ldots, \mathcal{I}_K$; we now want to show that, conversely, every function i, which depends rationally on $\mathcal{I}; \mathcal{I}_1, \mathcal{I}_2, \ldots, \mathcal{I}_K$ and integrally and algebraically on $\mathcal{I}_1, \mathcal{I}_2, \ldots, \mathcal{I}_K$, necessarily has to be an invariant polynomial. This is because such a function satisfies an equation of the form

$$i^k + \mathcal{F}_1(\mathcal{I}_1, \ldots, \mathcal{I}_K)\, i^{k-1} + \cdots + \mathcal{F}_k(\mathcal{I}_1, \ldots, \mathcal{I}_K) = 0,$$

where $\mathcal{F}_1, \mathcal{F}_2, \ldots, \mathcal{F}_k$ are integral rational functions of $\mathcal{I}_1, \mathcal{I}_2, \ldots, \mathcal{I}_K$. It can, furthermore, be expressed in the form

$$i = \frac{g}{h},$$

where g and h are integral rational functions in the coefficients of the base form that have no common factors. We, therefore, have that

$$\frac{g^k}{h^k} + \mathcal{F}_1 \frac{g^{k-1}}{h^{k-1}} + \mathcal{F}_2 \frac{g^{k-2}}{h^{k-2}} + \cdots + \mathcal{F}_k = 0,$$

or

$$\frac{g^k}{h} + \mathcal{F}_1 g^{k-1} + \mathcal{F}_2 g^{k-2} h + \cdots + \mathcal{F}_k h^{k-1} = 0,$$

or

$$\frac{g^k}{h} = -\mathcal{F}_1 g^{k-1} - \mathcal{F}_2 g^{k-2} h - \cdots - \mathcal{F}_k h^{k-1},$$

that is, g^k/h has to be an integral rational function of the coefficients, and since g and h have no common factors, h must be a constant. Hence, i must be an integral rational function of the coefficients of the base form, that is, an integral rational invariant. Thus, we have the following:

Theorem *The integral algebraic functions in the invariant field determined by $\mathcal{I}; \mathcal{I}_1, \mathcal{I}_2, \ldots, \mathcal{I}_K$ form precisely the system of all integral rational invariants.**

According to a fundamental theorem of Kronecker,** each function field contains a finite number of integral algebraic functions $\omega_1, \omega_2, \ldots, \omega_h$ such that every other integral algebraic function ω in the field can be brought into the form

$$\omega = \alpha_1\omega_1 + \alpha_2\omega_2 + \cdots + \alpha_h\omega_h,$$

where $\alpha_1, \ldots, \alpha_h$ are integral rational functions of $\mathcal{I}; \mathcal{I}_1, \mathcal{I}_2, \ldots, \mathcal{I}_K$. Thus, if we can find these elements $\omega_1, \omega_2, \ldots, \omega_h$, then the full invariant system is formed by $\omega_1, \omega_2, \ldots, \omega_h; \mathcal{I}; \mathcal{I}_1, \mathcal{I}_2, \ldots, \mathcal{I}_K$.

Now, every integral algebraic function ω of the field, which according to the above satisfies an equation of the form

$$\omega^k + \mathcal{F}_1(\mathcal{I}_1, \ldots, \mathcal{I}_K)\,\omega^{k-1} + \cdots + \mathcal{F}_k(\mathcal{I}_1, \ldots, \mathcal{I}_K) = 0,$$

can be expressed in the form

$$\omega = c_1\mathcal{I}^{k-1} + c_2\mathcal{I}^{k-2} + \cdots + c_k,$$

where $c_1, c_2, \ldots, c_k$ are rational functions of $\mathcal{I}_1, \mathcal{I}_2, \ldots, \mathcal{I}_K$. We know further that, if ω' is another root of the equation for ω, then we must necessarily have

$$\omega' = c_1{\mathcal{I}'}^{k-1} + c_2{\mathcal{I}'}^{k-2} + \cdots + c_k,$$

where $\mathcal{I}'$ is likewise a second root of the equation satisfied by $\mathcal{I}$, that is, the equation

$$\mathcal{I}^k + \mathcal{G}_1(\mathcal{I}_1, \ldots, \mathcal{I}_K)\,\mathcal{I}^{k-1} + \cdots + \mathcal{G}_k(\mathcal{I}_1, \ldots, \mathcal{I}_K) = 0.$$

Analogously, one has further that

$$\omega'' = c_1\mathcal{I}''^{k-1} + c_2\mathcal{I}''^{k-2} + \cdots + c_k,$$

$$\cdots$$

$$\omega^{(k-1)} = c_1{\mathcal{I}^{(k-1)}}^{k-1} + c_2{\mathcal{I}^{(k-1)}}^{k-2} + \cdots + c_k.$$

* In modern notation, the theorem states that $\mathbf{C}[V]^G = \mathbf{C}(V)^G \cap \mathbf{C}[V]$.
** See, e.g., Zariski and Samuel, *Commutative Algebra*, vol. I, 1958, p. 266.

These equations allow us to obtain information about the quantities $c_1, \ldots, c_k$. Solving them with respect to $c_1, c_2, \ldots, c_k$ gives:

$$c_1 = \pm \frac{\begin{vmatrix} 1 & \mathcal{I} & \mathcal{I}^2 & \cdots & \omega \\ 1 & \mathcal{I}' & \mathcal{I}'^2 & \cdots & \omega' \\ & & \cdots & & \\ 1 & \mathcal{I}^{(k-1)} & \mathcal{I}^{(k-1)^2} & \cdots & \omega^{(k-1)} \end{vmatrix}}{\begin{vmatrix} 1 & \mathcal{I} & \mathcal{I}^2 & \cdots & \mathcal{I}^{k-1} \\ & & \cdots & & \\ 1 & \mathcal{I}^{(k-1)} & \mathcal{I}^{(k-1)^2} & \cdots & \mathcal{I}^{(k-1)^{k-1}} \end{vmatrix}},$$

and analogously for the others. If we multiply numerator and denominator by the denominator, then we obtain in the denominator the discriminant of the equation satisfied by the $\mathcal{I}, \mathcal{I}', \ldots$; thus, setting this discriminant equal to D, we have:

$$c_1 = \pm \frac{Z}{D},$$

and so also

$$Z = \pm c_1 D.$$

But D depends integrally and rationally on the coefficients of the equation $\mathcal{I}^k + \mathcal{G}_1 \mathcal{I}^{k-1} + \cdots = 0$, hence D is also an integral rational function of $\mathcal{I}_1, \mathcal{I}_2, \ldots, \mathcal{I}_K$. Therefore, Z is a rational function of $\mathcal{I}_1, \mathcal{I}_2, \ldots, \mathcal{I}_K$. But since the elements of the two determinants in Z, which is the product of numerator and denominator in the above expression for c_1, are all integral algebraic functions of the $\mathcal{I}_1, \mathcal{I}_2, \ldots, \mathcal{I}_K$, it follows that Z is also an integral algebraic function of these quantities. And so, we finally conclude that Z is an integral rational function of $\mathcal{I}_1, \mathcal{I}_2, \ldots, \mathcal{I}_K$, since the quantities $\mathcal{I}_1, \mathcal{I}_2, \ldots, \mathcal{I}_K$ were determined in such a way that they do not satisfy any algebraic identity, together with the theorem that a function of certain mutually independent quantities, which is rational and at the same time integral and algebraic, has to be an integral rational function. An analogous conclusion holds for the other coefficients c_i, and it follows herewith that

$$\begin{aligned} \omega &= \frac{Z_1 \mathcal{I}^{k-1} + Z_2 \mathcal{I}^{k-2} + \cdots + Z_k}{D} \\ &= \frac{\mathcal{G}(\mathcal{I}; \mathcal{I}_1, \ldots, \mathcal{I}_K)}{D}; \end{aligned}$$

that is, one has the following:

Theorem *Every integral algebraic function of the invariant field, that is, every invariant of the base forms, is an integral rational function of $\mathcal{I}$; $\mathcal{I}_1, \ldots \mathcal{I}_K$, up to the discriminant of the equation defining the invariant field.*

Hence, to determine all invariants one only needs to determine the integral rational functions of $\mathcal{I}$; $\mathcal{I}_1, \ldots, \mathcal{I}_K$ that are divisible by D.

The system of invariants $\mathcal{I}_1, \mathcal{I}_2, \ldots, \mathcal{I}_K$, whose importance we saw in this section, possesses more fundamental properties, which we need to discuss in the following section.

Lecture XXXIX (July 16, 1897)

II.4 The vanishing of the invariants

Among all base forms, one can consider in particular those whose coefficients have numerical values such that the K invariants $\mathcal{I}_1, \mathcal{I}_2, \ldots, \mathcal{I}_K$ are all equal to zero. Since every other invariant is an integral algebraic function of these invariants, that is, satisfies an equation of the form

$$i^k + \mathcal{F}_1 i^{k-1} + \cdots + \mathcal{F}_k = 0,$$

where $\mathcal{F}_1, \mathcal{F}_2, \ldots, \mathcal{F}_k$ are integral rational functions of $\mathcal{I}_1, \ldots, \mathcal{I}_K$, which cannot contain a constant term since otherwise the homogeneity of the equation would be disturbed, it follows that in this case every invariant must vanish, because one has

$$i^k = 0,$$

from which it follows that $i = 0$. Such a base form, all of whose invariants vanish, is called a *null form.* We will soon realize the importance of studying the vanishing of the invariants. In any case, to begin with, we have the following:

Theorem *If the coefficients of the base form are given special values such that the K invariants $\mathcal{I}_1, \mathcal{I}_2, \ldots, \mathcal{I}_K$ become equal to zero, then all other invariants of the base form vanish also, that is, it is a null form.*

The converse of this theorem is true also, which is essential. We will show that if one knows of μ invariants $\mathcal{I}_1, \mathcal{I}_2, \ldots, \mathcal{I}_\mu$ of the base form such that their vanishing implies the vanishing of all other invariants of the base form, then all invariants have to be integral algebraic functions of $\mathcal{I}_1, \mathcal{I}_2, \ldots, \mathcal{I}_\mu$. This theorem is not entirely easy to prove; the proof will use the following general theorem.

Theorem *Suppose given* m *integral rational homogeneous functions* $f_1, f_2, \ldots, f_m$ *in the* n *variables* $x_1, x_2, \ldots, x_n$, *and let* $\mathcal{F}, \mathcal{F}', \mathcal{F}'', \ldots$ *be arbitrary integral rational homogeneous functions in the same variables* $x_1, \ldots, x_n$ *that vanish for all those values of the variables for which the given* m *functions* $f_1, f_2, \ldots, f_m$ *all vanish. Then one can always determine an integer* r *such that each product* $\prod_{(r)}$ *of* r *arbitrary functions of the sequence* $\mathcal{F}, \mathcal{F}', \mathcal{F}'', \ldots$ *can be expressed in the form*

$$\prod_{(r)} = a_1 f_1 + a_2 f_2 + \cdots + a_m f_m,$$

where $a_1, a_2, \ldots, a_m$ *are suitably chosen integral rational homogeneous functions in the variables* $x_1, x_2, \ldots, x_n$.*

This theorem is very important and applicable. It is very likely that it can be strengthened number theoretically. That is, if one assumes the coefficients of $f_1, \ldots, f_m$, $\mathcal{F}, \mathcal{F}', \ldots$ to be integral, then one can require $a_1, a_2, \ldots, a_m$ to be forms with integral coefficients. But a proof for this has not yet been given. Requiring integral coefficients is generally a sharpening of all theorems in this part whose possibility has yet to be proven.**

The proof of this theorem is very cumbersome, and it would lead us too far afield to discuss a proof here. (It can be found in Hilbert (1893).) But we would like to briefly demonstrate its significance through two examples.

We choose $n = 2$. That is, we consider m binary forms $f_1, \ldots, f_m$. These can all have a common factor, say ϕ. Then

$$f_1 = \phi \cdot \phi_1,$$
$$f_2 = \phi \cdot \phi_2,$$
$$\cdots$$
$$f_m = \phi \cdot \phi_m,$$

so that $\phi_1, \phi_2, \ldots, \phi_m$ do not have any more common factors. Then there are no more sets of values x_1, x_2 for which all forms $\phi_1, \ldots, \phi_m$ vanish.

* This theorem is now known as the Hilbert Nullstellensatz.

** See the following related papers: C. A. Berenstein and A. Yger, Effective Bezout identities in $\mathbf{Q}[z_1, \ldots, z_n]$, *Acta Math.* **166** (1991), 69–120, and P. Philippon, Denominateurs dans le théorème des zéros de Hilbert, *Acta Arithmetica* **58** (1991), 1–25.

Our theorem then says that every binary form whose order exceeds a certain number can be expressed linearly in terms of $\phi_1, \phi_2, \ldots, \phi_m$. We set $x_1 = x, x_2 = 1$, which gives us two integral rational functions in one variable x. As is well known, there is then a theorem that says that one can find two integral rational functions $a(x)$, $b(x)$ such that

$$1 = a(x)\phi_1(x) + b(x)\phi_2(x),$$

since ϕ_1 and ϕ_2 have no common factor. If we again set $x = \frac{x_1}{x_2}$, then it follows that

$$x_2^\nu = a \cdot \phi_1 + b \cdot \phi_2,$$

and, analogously,

$$x_1^\nu = a' \cdot \phi_1 + b' \cdot \phi_2,$$

so, indeed, each term $cx_1^\alpha x_2^\beta$, where $\alpha + \beta \geq 2\nu$, can be expressed in the desired fashion. Returning to the forms $f_1, f_2, \ldots, f_m$, one sees that a form $\mathcal{F}$ with the same zeroes as $f_1, \ldots, f_m$ has to be divisible by ϕ; thus, this problem is reduced to the previous one, because $\frac{\mathcal{F}}{\phi}$ can indeed be expressed in the desired fashion through $\phi_1, \ldots, \phi_m$, and hence $\mathcal{F}$ through $f_1, \ldots, f_m$.

For ternary forms, the proof can still be carried out by the previous line of reasoning. Given two ternary forms whose common zeroes are all distinct, the number $r = 1$ is already sufficient. One then has the theorem, which was first stated and proven by M. Noether, and which is thus only a special case of the general theorem:

Theorem *Given two algebraic curves $f_1^{(\alpha)} = 0$, $f_2^{(\beta)} = 0$, of orders α, β, respectively, such that the curves intersect in $\alpha \cdot \beta$ distinct points, every curve $\mathcal{F} = 0$ which also passes through these $\alpha \cdot \beta$ points can be expressed in the form:*

$$\mathcal{F} = a_1 f_1 + a_2 f_2 = 0,$$

where a_1 and a_2 are again ternary forms.

For three quaternary forms the proof is still not difficult; the general proof, however, is very cumbersome. It can be found in the cited work of Hilbert (1893).

Once this theorem is proven, the proof of the aforementioned theorem is not far. We now prove the following:

Theorem *If μ invariants $\mathcal{I}_1, \mathcal{I}_2, \ldots, \mathcal{I}_\mu$ have the property that their vanishing always implies necessarily the vanishing of all other invariants* *of the base form, then all invariants are integral algebraic functions of those* μ *invariants* $\mathcal{I}_1, \mathcal{I}_2, \ldots, \mathcal{I}_\mu$.

Lecture XL (July 19, 1897)

According to the general theorem one can find a number r such that every product $\prod_{(r)}$ of r or more arbitrary invariants of the base form can be written in the form

$$\prod_{(r)} = \mathcal{A}_1\mathcal{I}_1 + \mathcal{A}_2\mathcal{I}_2 + \cdots + \mathcal{A}_\mu\mathcal{I}_\mu.$$

Let $i_1, i_2, \ldots, i_m$ be a full invariant system; let the highest degree that appears in them be ν. Then each invariant $\mathcal{I}$ whose degree is $\geq r\nu$ can be written as a sum of products $\prod_{(r)}$ of r or more invariants i_k, since such a degree can only be reached if at least r factors appear. Thus, we have

$$\mathcal{I}_{(\geq r\nu)} = \mathcal{A}_1\mathcal{I}_1 + \mathcal{A}_2\mathcal{I}_2 + \cdots + \mathcal{A}_\mu\mathcal{I}_\mu.$$

But, as we already explained in Section II.2, the $\mathcal{A}_1, \mathcal{A}_2, \ldots, \mathcal{A}_\mu$ can be chosen to be invariants (Lecture XXXVI); hence

$$\mathcal{I}_{(\geq r\nu)} = \mathcal{J}_1\mathcal{I}_1 + \mathcal{J}_2\mathcal{I}_2 + \cdots + \mathcal{J}_\mu\mathcal{I}_\mu,$$

where now the invariants $\mathcal{J}$ have a lower degree than $\mathcal{I}$. Now, let $j_1, j_2, \ldots, j_N$ be all linearly independent invariants of degree $< \nu r$. Just as $\mathcal{I}$, one can represent $\mathcal{J}_1, \ldots, \mathcal{J}_\mu$ as linear functions of $\mathcal{I}_1, \ldots, \mathcal{I}_\mu$ whose coefficients are invariants, and this procedure can be continued until the coefficients become the invariants $j_1, j_2, \ldots, j_N$. Hence, one can put $\mathcal{I}$ into the form

$$\mathcal{I} = \mathcal{G}_1(\mathcal{I}_1, \ldots, \mathcal{I}_\mu)\, j_1 + \mathcal{G}_2(\mathcal{I}_1, \ldots, \mathcal{I}_\mu)\, j_2 + \cdots + \mathcal{G}_N(\mathcal{I}_1, \ldots, \mathcal{I}_\mu)\, j_N.$$

Furthermore, every invariant can be put into this form, because we have just proven it for degree $\geq r\nu$, and for degree $< r\nu$ it is self-evident because the latter are the invariants j themselves. The same is true, however, for the invariants $j_1\mathcal{I}, j_2\mathcal{I}, \ldots$, so we have

$$j_1\mathcal{I} = \mathcal{G}_1'j_1 + \mathcal{G}_2'j_2 + \cdots + \mathcal{G}_N'j_N,$$
$$j_2\mathcal{I} = \mathcal{G}_1''j_1 + \mathcal{G}_2''j_2 + \cdots + \mathcal{G}_N''j_N,$$
$$\cdots$$
$$j_N\mathcal{I} = \mathcal{G}_1^{(N)}j_1 + \mathcal{G}_2^{(N)}j_2 + \cdots + \mathcal{G}_N^{(N)}j_N,$$

for which we can also write

$$\begin{aligned} (\mathcal{G}_1' - \mathcal{I})\, j_1 + \mathcal{G}_2' j_2 + \cdots + \mathcal{G}_N' j_N &= 0, \\ \mathcal{G}_1'' j_1 + (\mathcal{G}_2'' - \mathcal{I})\, j_2 + \cdots + \mathcal{G}_N'' j_N &= 0, \\ &\cdots \\ \mathcal{G}_1^{(N)} j_1 + \mathcal{G}_2^{(N)} j_2 + \cdots + \left(\mathcal{G}_N^{(N)} - \mathcal{I}\right) j_N &= 0. \end{aligned}$$

Here, all $\mathcal{G}$ denote integral rational functions of the invariants $\mathcal{I}_1, \mathcal{I}_2, \ldots, \mathcal{I}_\mu$. If one now eliminates $j_1, j_2, \ldots, j_N$ from these equations, then it follows that

$$\begin{vmatrix} \mathcal{G}_1' - \mathcal{I} & \mathcal{G}_2' & \cdots & \mathcal{G}_N' \\ \mathcal{G}_1'' & \mathcal{G}_2'' - \mathcal{I} & \cdots & \mathcal{G}_N'' \\ & \cdots & & \\ \mathcal{G}_1^{(N)} & \mathcal{G}_2^{(N)} & \cdots & \mathcal{G}_N^{(N)} - \mathcal{I} \end{vmatrix} = 0,$$

and this equation proves that $\mathcal{I}$ is an integral algebraic function of $\mathcal{I}_1, \mathcal{I}_2, \ldots, \mathcal{I}_\mu$; thus, our theorem is proven. We remark that the theorem is not valid for arbitrary integral rational functions, since the invariant property of $\mathcal{I}, \mathcal{I}_1, \ldots, \mathcal{I}_\mu$ was used several times in an essential way.

This theorem is of the utmost importance and is very fruitful. We will see this immediately, because we will show in a series of examples how it can be used, among other things, for the determination of the full invariant system in a very convenient and transparent fashion.

In view of our theorem it is very important for the study of the invariants of a base form to know necessary and sufficient conditions for all invariants of this base form to vanish; if we interpret the N coefficients of the base form in the usual way as the coordinates of a projective space of $N-1$ dimensions, we are thereby led to the study of that algebraic object Z in this space, which is determined by setting all invariants equal to zero. If, as before, K denotes the number of algebraically independent invariants, then according to the earlier observations there are exactly K invariants $\mathcal{I}_1, \mathcal{I}_2, \ldots, \mathcal{I}_K$ whose vanishing already completely determines the algebraic object Z; from the theorem just proved it then follows that necessarily $\mu \geq K$, that is, it is not possible to find a smaller number of invariants whose zeroes already determine the object Z.

We explain these observations, and apply our theorem to binary forms. For a quadratic form, the vanishing of the only existing invariant says that the form has a double factor, that is, if for

$$f = a_0 x_1^2 + 2a_1 x_1 x_2 + a_2 x_2^2$$

we have

$$D = a_0 a_2 - a_1^2 = 0,$$

then f can be linearly transformed into

$$f' = a x_1^2.$$

Likewise, the cubic form

$$f = a_0 x_1^3 + 3a_1 x_1^2 x_2 + 3a_2 x_1 x_2^2 + a_3 x_2^3$$

can be linearly transformed into

$$f' = x_1^2 (m x_1 + n x_2),$$

if its only invariant (the discriminant) is $D = 0$. If the two invariants i, j of the biquadratic form

$$f = a_0 x_1^4 + \cdots + a_4 x_2^4$$

vanish, that is, if

$$i = 0,\ j = 0,$$

then one can, since f then has a threefold factor, transform the form linearly into

$$f' = x_1^3 (m x_1 + n x_2).$$

One can proceed analogously for forms of even higher order.

We now use these facts, proven earlier (Section I.8), to establish the full system of in- and covariants for single forms, as well as systems of forms, for which our theorem serves as a very useful tool.

Instead of considering the in- and covariants of a form, we can consider the invariants of two simultaneous forms, which we need to do since our theorem applies only to invariants. But since one is justified to talk only of invariants instead of covariants, one realizes that our methods are indeed quite general. First, we want to construct the full form system of the binary cubic form, so we consider the two forms:

$$f = a_0 x_1^3 + 3a_1 x_1^2 x_2 + 3a_2 x_1 x_2^2 + a_3 x_2^3,$$
$$g = b_0 x_1 + b_1 x_2.$$

If we indicate by bars the forms we obtain by replacing x_1 by b_1, x_2

by $-b_0$, then, according to earlier observations we obtain the following simultaneous invariants of f and g:

$$\overline{f} = f(b_1, -b_0) = a_0 b_1^3 - 3a_1 b_1^2 b_0 + \cdots,$$
$$\overline{\mathcal{H}} = \left(a_0 a_2 - a_1^2\right) b_1^2 - (a_0 a_3 - a_1 a_2)\, b_0 b_1 + \left(a_1 a_3 - a_2^2\right) b_0^2,$$
$$\mathcal{D} = a_0^2 a_3^2 + \cdots,$$
$$\overline{\mathcal{F}} = \left(a_0^2 a_3 + \cdots\right) b_1^3 + \cdots.$$

Between them exists the syzygy

$$4\overline{\mathcal{H}}^3 = \mathcal{D}\overline{f}^2 - \overline{\mathcal{J}}^2.$$

Lecture XLI (July 20, 1897)

We have seen earlier that all invariants can be expressed rationally in terms of $\overline{f}, \overline{\mathcal{H}}, \overline{\mathcal{J}}$. On the other hand, we now want to show that all invariants can be expressed integrally and algebraically in terms of $\overline{f}, \overline{\mathcal{H}}, \mathcal{D}$. If $\mathcal{D} = 0$, then f has a double factor; hence we can set $a_2 = 0$, $a_3 = 0$— assuming the application of the appropriate transformation. From $\overline{\mathcal{H}} = 0$ it follows that

$$a_1^2 b_1^2 = 0,$$

and from $\overline{f} = 0$, that

$$a_0 b_1^3 - 3a_1 b_1^2 b_0 = 0.$$

Hence, either $b_1 = 0$ or, if we want to assume $b_1 \neq 0$, then $a_1 = 0$, $a_0 = 0$. In the latter case, all invariants vanish, however, since g alone does not have any. So assume that

$$a_2 = 0,\ a_3 = 0,\ b_1 = 0.$$

Hence, the vanishing of the invariants $\overline{f}, \overline{\mathcal{H}}, \mathcal{D}$ says that the form g is contained twice in f. But from this one concludes easily that, consequently, all simultaneous invariants of f and g have to vanish. That is because these are of the form (Section I.9)

$$\sum Z a_0^{\nu_0} a_1^{\nu_1} a_2^{\nu_2} a_3^{\nu_3} b_0^{\mu_0} b_1^{\mu_1},$$

where

$$\nu_0 + \nu_1 + \nu_2 + \nu_3 = r,$$
$$\mu_0 + \mu_1 = s,$$
$$\nu_1 + 2\nu_2 + 3\nu_3 + \mu_1 = \frac{3r + s}{2}$$

are constants. In this invariant all terms containing a_2, a_3, or b_1 vanish. Thus, we only need to prove that those terms for which $\nu_2 = 0$, $\nu_3 = 0$, $\mu_1 = 0$ vanish too. For those we have

$$\nu_0 + \nu_1 = r,$$
$$\mu_0 = s,$$
$$\nu_1 = \frac{3r+s}{2},$$

whence

$$2\nu_1 = 3\nu_0 + 3\nu_1 + \mu_0$$

or

$$3\nu_0 + \nu_1 + \mu_0 = 0.$$

Since the exponents are positive integers, it follows that

$$\nu_0 = 0,\ \nu_1 = 0,\ \mu_0 = 0.$$

Thus, if the whole invariant is not to be a constant—which we never consider—these terms have to be zero, thus, don't exist at all. The vanishing of $\overline{\mathcal{f}}$, $\overline{\mathcal{H}}$, $\mathcal{D}$, therefore, implies the vanishing of all other invariants. According to our theorem, each invariant is, therefore, an integral algebraic function of $\overline{\mathcal{f}}$, $\overline{\mathcal{H}}$, $\mathcal{D}$, which is confirmed by the syzygy for the invariant $\overline{\mathcal{J}}$.

There cannot exist an algebraic relation between $\overline{\mathcal{f}}$, $\overline{\mathcal{H}}$, $\mathcal{D}$ since their number coincides with the number $4+2-3=3$ of constants* which necessarily occur in f and g. (One needs to subtract 3 from the total number since 3 constants can be eliminated by a linear transformation.) So, we can view $\overline{\mathcal{f}}$, $\overline{\mathcal{H}}$, $\mathcal{D}$ as the three independent variables. Hence, according to the theorem at the end of Section II.3, each invariant can be represented in the form

$$\omega = \frac{A + B\overline{\mathcal{J}}}{\mathcal{D}\overline{\mathcal{f}}^2 - 4\overline{\mathcal{H}}^3},$$

because we have to add the invariant $\overline{\mathcal{J}}$ to the invariants $\overline{\mathcal{f}}$, $\overline{\mathcal{H}}$, $\mathcal{D}$ in order to obtain rational representation. That is, we have to consider the field of the equation

$$\overline{\mathcal{J}}^2 - \mathcal{D}\overline{\mathcal{f}}^2 + 4\overline{\mathcal{H}}^3 = 0,$$

* The group $G = SL_2(\mathbf{C})$ acts transitively on $S^3\mathbf{C}^2 \oplus \mathbf{C}^2$, hence $\operatorname{tr}\deg_{\mathbf{C}} \mathbf{C}[V]^G \geq \dim(V) - \dim(G)$.

whose discriminant is $\mathcal{D}\overline{f}^2 - 4\overline{\mathcal{H}}^3$. Here, A and B are integral rational functions of $\mathcal{D}$, $\overline{f}$, $\overline{\mathcal{H}}$. But then the conjugated quantity

$$\omega' = \frac{A - B\overline{\mathcal{J}}}{\mathcal{D}\overline{f}^2 - 4\overline{\mathcal{H}}^3}$$

is also an integral algebraic function of the field, thus an invariant. And, since $\overline{\mathcal{J}}$ is not divisible by the denominator, but $\omega + \omega'$ and $\omega - \omega'$ have to be invariants also, it therefore follows that A, as well as B, has to be divisible by the denominator. Hence every invariant can be brought into the form

$$\omega = A_1 + B_1 \mathcal{J},$$

where A_1, B_1 are integral rational functions of $\overline{f}$, $\overline{\mathcal{H}}$, $\mathcal{D}$. So we have proven the following:

Theorem *The full system of forms of the binary cubic form is given by* f, $\mathcal{H}$, $\mathcal{J}$, $\mathcal{D}$. *Between these there exists exactly one syzygy.*

We use the exact same procedure to determine the full system of covariants for the binary biquadratic form. To this aim we consider the invariants of the two forms:

$$\begin{aligned} f &= a_0 x_1^4 + 4a_1 x_1^3 x_2 + 6a_2 x_1^2 x_2^2 + 4a_3 x_1 x_2^3 + a_4 x_2^4, \\ g &= b_0 x_1 + b_1 x_2. \end{aligned}$$

We certainly have the following invariants:

$$\begin{aligned} &\overline{f}, \\ i &= a_0 a_4 - 4a_1 a_3 + 3a_2^2, \\ j &= a_0 a_2 a_4 + \cdots, \\ \overline{\mathcal{J}} &= \left(a_0^2 a_3 - 3a_0 a_1 a_2 + 2a_1^3\right) b_1^6 + \cdots, \\ \overline{\mathcal{H}} &= \left(a_0 a_2 - a_1^2\right) b_1^4 + \cdots \\ &= \begin{vmatrix} a_0 b_1^2 - 2a_1 b_0 b_1 + a_2 b_0^2 & a_1 b_1^2 - 2a_2 b_0 b_1 + a_3 b_0^2 \\ a_1 b_1^2 - 2a_2 b_0 b_1 + a_3 b_0^2 & a_2 b_1^2 - 2a_3 b_0 b_1 + a_4 b_0^2 \end{vmatrix}. \end{aligned}$$

Between them exists the following syzygy (Section I.8)

$$\overline{\mathcal{J}}^2 + 4\overline{\mathcal{H}}^3 - i\overline{f}^2\overline{\mathcal{H}} + j\overline{f}^3 = 0.$$

According to Section I.8, all invariants can be expressed rationally in terms of $\overline{f}$, $\overline{\mathcal{H}}$, $\overline{\mathcal{J}}$, i. On the other hand, we will now show that every invariant can be expressed integrally and algebraically in terms of

i, j, $\overline{f}$, $\overline{\mathcal{H}}$, because, if i and j are equal to zero, then the discriminant is zero also. So one can set $a_3 = 0$, $a_4 = 0$; but we also have $a_2 = 0$ because $i = 0$; hence f has a threefold linear factor. Furthermore, since $\overline{\mathcal{H}} = 0$, we have

$$a_1^2 b_1^4 = 0,$$

and from $\overline{f} = 0$ it follows that

$$a_0 b_1^4 - 4a_1 b_1^3 b_0 = b_1^3 (a_0 b_1 - 4a_1 b_0) = 0.$$

If $b_1 \neq 0$, then $a_1 = 0$, $a_0 = 0$, so all invariants vanish. But it can also happen that $b_1 = 0$. These conditions,

$$a_2 = 0,\ a_3 = 0,\ a_4 = 0,\ b_1 = 0,$$

say that the form g occurs three times in f.

Lecture XLII (July 22, 1897)

Each invariant can now be represented in the form

$$\sum Z a_0^{\nu_0} a_1^{\nu_1} a_2^{\nu_2} a_3^{\nu_3} a_4^{\nu_4} b_0^{\mu_0} b_1^{\mu_1},$$

where

$$\nu_0 + \nu_1 + \nu_2 + \nu_3 + \nu_4 = r,$$
$$\mu_0 + \mu_1 = s,$$
$$\nu_1 + 2\nu_2 + 3\nu_3 + 4\nu_4 + \mu_1 = \frac{4r + s}{2}$$

are constants. To prove that the invariant vanishes we only need to show that there are no terms for which $\nu_2 = 0$, $\nu_3 = 0$, $\nu_4 = 0$, $\mu_1 = 0$, because for those terms

$$\nu_0 + \nu_1 = r,$$
$$\mu_0 = s,$$
$$\nu_1 = 2\nu_0 + 2\nu_1 + \frac{\mu_0}{2},$$

whence

$$4\nu_0 + 2\nu_1 + \mu_0 = 0.$$

Since ν_0, ν_1, μ_0 have to be nonnegative, it follows that

$$\nu_0 = 0,\ \nu_1 = 0,\ \mu_0 = 0.$$

Therefore, all invariants vanish necessarily if i, j, $\overline{f}$, $\overline{\mathcal{H}}$ vanish; hence, every invariant is an integral algebraic function of these. If one adds $\overline{\mathcal{J}}$ to them, then rational representation of each invariant is possible. Consequently, we have to consider the field $\overline{\mathcal{J}}$, i, j, $\overline{f}$, $\overline{\mathcal{H}}$, for which we have the equation

$$\overline{\mathcal{J}}^2 - i\overline{f}^2\overline{\mathcal{H}} + 4\overline{\mathcal{H}}^3 + j\overline{f}^3 = 0.$$

According to the theorem at the end of Section II.3, each invariant can be expressed in the form

$$\omega = \frac{A + B\overline{\mathcal{J}}}{i\overline{f}^2\overline{\mathcal{H}} - 4\overline{\mathcal{H}}^3 - j\overline{f}^3},$$

and, therefore, the conjugate expression

$$\omega' = \frac{A - B\overline{\mathcal{J}}}{i\overline{f}^2\overline{\mathcal{H}} - 4\overline{\mathcal{H}}^3 - j\overline{f}^3}$$

is also an invariant. From these two equations we conclude immediately that A and B are divisible by the denominator; hence, every invariant can be brought into the form

$$\omega = A_1 + b_1\overline{\mathcal{J}},$$

where A_1, B_1 are integral rational functions of i, j, $\overline{f}$, $\overline{\mathcal{H}}$. Thus, we have the following:

Theorem *The full form system of the biquadratic form is given by the forms f, $\mathcal{H}$, $\mathcal{J}$, i, j. Between them there exists a unique syzygy.*

One proceeds completely analogously in the construction of the full form system for simultaneous forms. We consider the simultaneous in- and covariants of a quadratic and a linear binary form, that is, the simultaneous invariants of a quadratic and two linear forms:

$$\begin{aligned} f &= a_0x_1^2 + 2a_1x_1x_2 + a_2x_2^2, \\ g &= b_0x_1 + b_1x_2, \\ l &= c_0x_1 + c_1x_2, \end{aligned}$$

where l denotes the added form. We already know the following simultaneous invariants for them (compare Section I.9):

$$\begin{aligned} &\overline{f}, \overline{g}, \\ h &= a_0a_2 - a_1^2, \\ r &= a_0b_1^2 - 2a_1b_0b_1 + a_2b_0^2, \\ \overline{\mathcal{I}} &= (a_0b_1 - a_1b_0)\,c_1 - (a_1b_1 - a_2b_0)\,c_0. \end{aligned}$$

We also have seen already that all other invariants can be expressed rationally through $\overline{f}$, $\overline{g}$, h, $\overline{\mathcal{I}}$; and that, furthermore, we obtained an expression for r in terms of the following syzygy:

$$\overline{\mathcal{I}}^2 = r\overline{f} - h\overline{g}^2.$$

On the other hand, we now show that all invariants can be expressed integrally and algebraically in terms of $\overline{f}$, $\overline{g}$, r, h. Namely, the vanishing of h implies that f can be put in the form

$$f = a_0 x_1^2,$$

hence that one can set $a_1 = 0$, $a_2 = 0$. The vanishing of $\overline{f}$, r, $\overline{g}$ then implies that

$$\begin{aligned} a_0 c_1^2 &= 0, \\ a_0 b_1^2 &= 0, \\ b_0 c_1 - b_1 c_0 &= 0. \end{aligned}$$

If $a_0 = 0$, $b_0 c_1 - b_1 c_0 = 0$, then all invariants containing a vanish; but then the two forms g, l have only one invariant $b_0 c_1 - b_1 c_0$, which then also vanishes. If $a_0 \neq 0$, then $c_1 = 0$, $b_1 = 0$; hence one has

$$a_1 = 0,\ a_2 = 0,\ c_1 = 0,\ b_1 = 0,$$

that is, the two linear forms g, l are in essence identical, and at the same time identical with the linear form that appears twice in f. Every invariant has the form:

$$\sum Z a_0^{\nu_0} a_1^{\nu_1} a_2^{\nu_2} b_0^{\mu_0} b_1^{\mu_1} c_0^{\rho_0} c_1^{\rho_1},$$

where

$$\begin{aligned} \nu_0 + \nu_1 + \nu_2 &= r, \\ \mu_0 + \mu_1 &= s, \\ \rho_0 + \rho_1 &= t, \\ \nu_1 + 2\nu_2 + \mu_1 + \rho_1 &= \frac{2r + s + t}{2} \end{aligned}$$

are constants. Only those terms need to be considered for which $\nu_1 = 0$, $\nu_2 = 0$, $\mu_1 = 0$, $\rho_1 = 0$. For those we have

$$\begin{gathered} \nu_0 = r,\ \mu_0 = s,\ \rho_0 = t, \\ 0 = 2\nu_0 + \mu_0 + \rho_0, \end{gathered}$$

whence $\nu_0 = 0$, $\mu_0 = 0$, $\rho_0 = 0$. Hence, such terms do not even occur, and the invariant is therefore zero if $\overline{f} = 0$, $\overline{g} = 0$, $r = 0$, $h = 0$. Thus, every invariant is an integral algebraic function of $\overline{f}$, $\overline{g}$, r, h. If we add $\overline{\mathcal{I}}$, then a rational representation is possible. We therefore have to consider the field formed by the invariants $\overline{\mathcal{I}}$, $\overline{f}$, $\overline{g}$, r, h for which the equation

$$\overline{\mathcal{I}}^2 - r\overline{f} + h\overline{g}^2 = 0$$

holds. Each invariant can be given the form

$$\omega = \frac{A + B\overline{\mathcal{I}}}{r\overline{f} - h\overline{g}^2},$$

and the conjugate expression

$$\omega' = \frac{A - B\overline{\mathcal{I}}}{r\overline{f} - h\overline{g}^2}$$

must also be an invariant. Since, therefore, A and B have each to be divisible by $r\overline{f} - h\overline{g}^2$, it follows that we can set

$$\omega = A_1 + B_1\overline{\mathcal{I}},$$

where A_1, B_1 are integral rational functions of $\overline{f}$, $\overline{g}$, h, r. Hence, we have the following:

Theorem *The full form system of a quadratic and a linear form is given by the forms* f, g, h, r, $\mathcal{I}$.

The same line of reasoning can also be applied without change in the case of the full form system of two quadratic binary forms, that is, the full form system of one linear and two quadratic forms. Let those forms be

$$\begin{aligned} f &= a_0x_1^2 + 2a_1x_1x_2 + a_2x_2^2, \\ g &= b_0x_1^2 + 2b_1x_1x_2 + b_2x_2^2, \\ l &= c_0x_1 + c_1x_2. \end{aligned}$$

These have, as we saw earlier (Section I.9), the following simultaneous invariants:

$$\begin{aligned} &\overline{f}, \overline{g}, \\ &\mathcal{H} = a_0a_2 - a_1^2, \\ &H = b_0b_2 - b_1^2, \\ &\overline{s}_1 = \overline{(f, g)}_1, \\ &s_2 = (f, g)_2, \end{aligned}$$

and between them exists the following syzygy:

$$H\overline{f}^2 - s_2\overline{f}\overline{g} + \mathcal{H}\overline{g}^2 + \overline{s}_1^2 = 0.$$

We also saw that all invariants can be expressed rationally in terms of $\overline{f}$, $\mathcal{H}$, $\overline{g}, \overline{s}_1$, s_2. On the other hand, we now show that all invariants can be expressed integrally and algebraically in terms of $\overline{f}$, $\overline{g}$, $\mathcal{H}$, H, s_2. We observe what their vanishing implies. The discriminant of $f+\lambda g$, where λ is an arbitrary parameter, is

$$\mathcal{D}(f+\lambda g) = \mathcal{H} + \lambda s_2 + \lambda^2 H;$$

hence $\mathcal{D}(f+\lambda g) = 0$ for every λ. Therefore, $f = a_x^2$, $g = b_x^2$ must be squares; but, for instance, for $\lambda = -1$,

$$f - g = a_x^2 - b_x^2$$

must also be a square. Thus, the two factors $(a_x - b_x)$ and $(a_x + b_x)$ can only differ by a constant factor. Therefore,

$$a_x - b_x = m(a_x + b_x),$$

whence

$$a_x(1-m) = b_x(1+m),$$

so a_x and b_x are not essentially different from each other. The vanishing of $\mathcal{H}$, H, s_2 therefore says that f and g can be brought into the form

$$f = a_0 x_1^2,$$
$$g = b_0 x_1^2.$$

Furthermore, we should have

$$\overline{f} = a_0 c_1^2 = 0,$$
$$\overline{g} = b_0 c_1^2 = 0.$$

Either $c_1 \neq 0$, and then $a_0 = 0$, $b_0 = 0$, and all invariants vanish since l alone does not possess any. Or $c_1 = 0$ besides $a_1 = 0$, $a_2 = 0$; $b_1 = 0$, $b_2 = 0$, in which case the vanishing of $\overline{f}$, $\overline{g}$, $\mathcal{H}$, H, s_2 says that the linear form l occurs twice in f as well as in g. An invariant has the form

$$\sum Z a_0^{\nu_0} a_1^{\nu_1} a_2^{\nu_2} b_0^{\mu_0} b_1^{\mu_1} b_2^{\mu_2} c_0^{\rho_0} c_1^{\rho_1},$$

where

$$\nu_0 + \nu_1 + \nu_2 = r,$$
$$\mu_0 + \mu_1 + \mu_2 = s,$$
$$\rho_0 + \rho_1 = t,$$
$$\nu_1 + 2\nu_2 + \mu_1 + 2\mu_2 + \rho_1 = \frac{2r + 2s + t}{2}$$

are constants. Only those terms can occur for which

$$\nu_1 = 0,\ \nu_2 = 0;\ \mu_1 = 0,\ \mu_2 = 0;\ \rho_1 = 0.$$

For those one has

$$0 = 2\nu_0 + 2\mu_0 + \rho_0,$$

whence $\nu_0 = 0$, $\mu_0 = 0$, $\rho_0 = 0$. Thus, such terms cannot occur at all and, consequently, all invariants vanish if $\overline{f}$, $\overline{g}$, $\mathcal{H}$, H, s_2 vanish. Hence, each invariant is an integral algebraic function of $\overline{f}$, $\overline{g}$, $\mathcal{H}$, H, s_2, and if one adds $\overline{s}_1$, then a rational representation is possible. Because of the equation

$$\overline{s}_1^2 + H\overline{f}^2 - s_2\overline{f}\overline{g} + \mathcal{H}\overline{g}^2 = 0$$

every invariant is representable in the form:

$$\omega = \frac{A + B\overline{s}_1}{H\overline{f}^2 - s_2\overline{f}\overline{g} + \mathcal{H}\overline{g}^2}.$$

Then the conjugate expression

$$\omega' = \frac{A - B\overline{s}_1}{H\overline{f}^2 - s_2\overline{f}\overline{g} + \mathcal{H}\overline{g}^2}$$

is also an invariant, and since consequently A and B are divisible by the denominator, it follows that

$$\omega = A_1 + B_1\overline{s}_1,$$

where A_1, B_1 are integral rational functions of $\overline{f}$, $\overline{g}$, $\mathcal{H}$, H, s_2. Thus we have the following:

Theorem *The full form system of two simultaneous quadratic forms is given by* f, g, $\mathcal{H}$, H, s_1, s_2.

The procedure we have used here for the construction of the full invariant system is completely general. We will also illustrate the line of reasoning briefly with the example of the full invariant system of three quadratic forms, as well as that of two cubic forms.

Lecture XLIII (July 23, 1897)

We consider the joint invariants of the three quadratic forms

$$\begin{aligned} f &= a_0x_1^2 + 2a_1x_1x_2 + a_2x_2^2, \\ g &= b_0x_1^2 + 2b_1x_1x_2 + b_2x_2^2, \\ k &= c_0x_1^2 + 2c_1x_1x_2 + c_2x_2^2. \end{aligned}$$

According to earlier observations we have at least the following invariants:

$$\begin{aligned}\mathcal{D}_f &= a_0a_2 - a_1^2,\\ \mathcal{D}_g &= b_0b_2 - b_1^2,\\ \mathcal{D}_k &= c_0c_2 - c_1^2,\\ (f,g)_2 &= a_0b_2 - 2a_1b_1 + a_2b_0,\\ (f,k)_2 &= a_0c_2 - 2a_1c_1 + a_2c_0,\\ (g,k)_2 &= b_0c_2 - 2b_1c_1 + b_2c_0,\end{aligned}$$

$$d = \begin{vmatrix} a_0 & a_1 & a_2 \\ b_0 & b_1 & b_2 \\ c_0 & c_1 & c_2 \end{vmatrix}.$$

One can now show easily that all other invariants can be expressed integrally and algebraically in terms of $\mathcal{D}_f$, $\mathcal{D}_g$, $\mathcal{D}_k$, $(f,g)_2$, $(f,k)_2$, $(g,k)_2$. Because, if one forms the discriminant of the form $f + \gamma g + \kappa k$, where γ, κ are arbitrary quantities, then one finds

$$\mathcal{D}(f+\gamma g+\kappa k) = \mathcal{D}_f + \gamma(f,g)_2 + \kappa(f,k)_2 + \gamma^2\mathcal{D}_g + +\gamma\kappa(g,k)_2 + \kappa^2\mathcal{D}_k.$$

Thus, if the above six invariants are zero, then $\mathcal{D}(f+\gamma g+\kappa k)$ becomes zero for all values of γ, κ, from which it follows not only that f, g, k have to be squares of linear factors, but also that these three linear factors all have to be identical. To see this, one only needs to first set $\gamma = -1$, $\kappa = 0$, and then $\gamma = 0$, $\kappa = -1$. Then the assertion follows in the aforementioned way.

One can, therefore, transform the three forms so that they simultaneously take on the form:

$$\begin{aligned} f &= a_0x_1^2,\\ g &= b_0x_1^2,\\ k &= c_0x_1^2.\end{aligned}$$

Just as before, one now concludes easily that terms of an invariant that only contain a_0, b_0, c_0 cannot exist; hence, all invariants are zero. Therefore, every invariant is an integral algebraic function of $\mathcal{D}_f$, $\mathcal{D}_g$, $\mathcal{D}_k$, $(f,g)_2$, $(f,k)_2$, $(g,k)_2$. These must be independent of each other since their number equals the number of mutually independent constants in f, g, k, namely, $9-3=6$.

For instance, one sees easily that d is an integral algebraic function of those invariants. Apparently one has:

$$2d^2 = \begin{vmatrix} a_0 & a_1 & a_2 \\ b_0 & b_1 & b_2 \\ c_0 & c_1 & c_2 \end{vmatrix} \cdot \begin{vmatrix} a_2 & -2a_1 & a_0 \\ b_2 & -2b_1 & b_0 \\ c_2 & -2c_1 & c_0 \end{vmatrix}$$

$$= \begin{vmatrix} 2\mathcal{D}_f & (f,g)_2 & (f,k)_2 \\ (g,f)_2 & 2\mathcal{D}_g & (g,k)_2 \\ (k,f)_2 & (k,g)_2 & 2\mathcal{D}_k \end{vmatrix}.$$

On the other hand, one can show, the details of which we will not go into, that all invariants can be expressed rationally in terms of all seven invariants, so one needs to add d to the above system of six invariants. Hence, the last equation is the equation of the invariant field. But from this the conclusion is clear that all invariants can be expressed integrally and rationally in terms of the seven invariants d, $\mathcal{D}_f$, $\mathcal{D}_g$, $\mathcal{D}_k$, $(f,g)_2$, $(f,k)_2\,(g,k)_2$.

If one has two cubic forms

$$f = a_0x_1^3 + \cdots,$$

$$g = b_0x_1^3 + \cdots,$$

and wants to determine their full invariant system, then this can be done as follows. Form the discriminant of $f + \gamma g$, that is,

$$\mathcal{D}(f + \gamma g) = \mathcal{D}_f + \gamma\mathcal{D}_1 + \gamma^2\mathcal{D}_2 + \gamma^3\mathcal{D}_3 + \gamma^4\mathcal{D}_g.$$

The coefficients $\mathcal{D}_f$, $\mathcal{D}_1$, $\mathcal{D}_2$, $\mathcal{D}_3$, $\mathcal{D}_g$ must be invariants, since $\mathcal{D}(f+\gamma g)$ is an invariant for all γ. Through it one can express all other invariants integrally and algebraically. Since, if they are zero, then $f + \gamma g$ always has a double factor, and so f and g have a common double factor—otherwise one could find values for γ so that $f + \gamma g$ does not have a double factor—and we can set

$$a_2 = 0,\ a_3 = 0;\ b_2 = 0,\ b_3 = 0.$$

For a given invariant we then need to consider

$$\sum Z a_0^{\nu_0} a_1^{\nu_1} b_0^{\mu_0} b_1^{\mu_1},$$

where we must have

$$\nu_0 + \nu_1 = r,$$

$$\mu_0 + \mu_1 = s,$$

$$2\nu_1 + 2\mu_1 = 3\nu_0 + 3\nu_1 + 3\mu_0 + 2\mu_1,$$

from which it follows that $\nu_0 = 0$, $\nu_1 = 0$, $\mu_0 = 0$, $\mu_1 = 0$. Hence, all invariants can be expressed integrally and algebraically through those five, which are independent of each other.

One would now have to look for an invariant which, added to those five, allows rational representation. Subsequently, one would have to establish all mutually independent integral algebraic functions of the so determined function field, a task for which the often-used theorem at the end of Section II.3 is of great service. If the problem is approached in this way, one finds that one only needs to add two invariants to the five to obtain the full invariant system, namely, the two invariants $(f,g)_3$ and $\mathcal{R}$, where $(f,g)_3$ is the third transvection, and $\mathcal{R}$ is the resultant of the two cubic forms. A calculation convinces one that the two latter are integral algebraic functions of those five invariants.

Finally, for two biquadratic forms

$$f = a_0 x_1^4 + \cdots,$$
$$g = b_0 x_1^4 + \cdots$$

one proceeds entirely analogously. One forms

$$i(f+\gamma g) = i_f + \gamma i_1 + \gamma^2 i_g,$$
$$j(f+\gamma g) = j_f + \gamma j_1 + \gamma^2 j_2 + \gamma^3 j_g.$$

All other invariants can be expressed through the seven invariants i_f, i_g, i_1, j_f, j_g, j_1, j_2, since their simultaneous vanishing says that f and g both contain the same linear factor thrice. At the same time, they are independent of each other. If one then adds another invariant which allows a rational representation, then one realizes that the eight invariants found in this way form the full invariant system of two biquadratic forms.

These examples should illustrate the method sufficiently. At the same time, the question arises here as to the degree of the invariant field; since the difficulty of determining the full invariant system depends, among other things, on the degree.

For a binary base form, Prof. Hilbert has found a formula for this degree, and it is very probable that this formula can be generalized in an appropriate way to a simultaneous system of binary forms.

Lecture XLIV (July 26, 1897)

From the observations of the past sections the significance of the null forms becomes apparent, and one recognizes, in particular, how important it is to know what the vanishing of all invariants says about the base forms, that is, to recognize the real meaning of the null forms. Thus, the essential question arises, whether the null forms cannot be characterized through other particularly simple properties. For binary forms this can indeed be done in a particularly simple way, as is shown by the following:

Theorem *If all invariants of a binary base form of order $n = 2h+1$, respectively $n = 2h$, are zero, then the base form possesses an $(h+1)$-fold linear factor—and conversely, if it possesses an $(h+1)$-fold linear factor, then all invariants are equal to zero.*

This theorem says that every null form can be put into the form:

$$f = x_1^{h+1}\left(a_0 x_1^{n-h-1} + \binom{n}{1} a_1 x_1^{n-h-2} x_2 + \cdots \right.$$
$$\left. + a_{n-h-1} x_2^{n-h-1} \left(\begin{matrix} n \\ n-h-1 \end{matrix}\right)\right),$$

and that, conversely, the latter is always a null form. Because of this, this last form is also called the *canonical null form.* Thus, a null form contains $n-h$ arbitrary parameters; the latter determine all coefficients.

We first prove the first part of the theorem, and assume that all invariants are equal to zero. We form the h transvections of the form

$$f = a_0 x_1^n + \binom{n}{1} a_1 x_1^{n-1} x_2 + \cdots + a_n x_2^n$$

over itself, that is, the second, fourth, ..., $2h$-fold transvections, which are

$$\mathcal{F}_1 = \left(a_0 a_2 - a_1^2\right) x_1^{2(n-2)} + \cdots,$$
$$\mathcal{F}_2 = \left(a_0 a_4 - 4a_1 a_3 + 3a_2^2\right) x_1^{2(n-4)} + \cdots,$$
$$\mathcal{F}_3 = \left(a_0 a_6 - 6a_1 a_5 + 15a_2 a_4 - 1 - a_3^2\right) x_1^{2(n-6)} + \cdots,$$
$$\cdots$$
$$\mathcal{F}_h = \left(a_0 a_{2h} - \binom{2h}{1} a_1 a_{2h-1} + \cdots \pm \frac{1}{2}\binom{2h}{h} a_2^2\right) x_1^{2(n-2h)} + \cdots.$$

From these covariants we form new ones that all have the same order. Let M be the least common multiple of the numbers n, $2(n-2)$, $2(n-4)$, $\ldots, 2(n-2h)$, respectively $2(n-2h+2)$, and set

$$M = mn = 2m_1(n-2) = 2m_2(n-4) = \cdots = 2m_h(n-2h),$$

respectively

$$M = mn = 2m_1(n-2) = 2m_2(n-4) = \cdots = 2m_{h-1}(n-2h+2),$$

depending on whether $n = 2h+1$ or $n = 2h$. Then we consider the two forms

$$\mathcal{U} = uf^m + u_1\mathcal{F}_1^{m_1} + u_2\mathcal{F}_2^{m_2} + \cdots + u_h\mathcal{F}_h^{m_h},$$
$$\mathcal{V} = vf^m + v_1\mathcal{F}_1^{m_1} + v_2\mathcal{F}_2^{m_2} + \cdots + v_h\mathcal{F}_h^{m_h},$$

respectively

$$\mathcal{U} = uf^m + u_1\mathcal{F}_1^{m_1} + u_2\mathcal{F}_2^{m_2} + \cdots + u_{h-1}\mathcal{F}_{h-1}{}^{m_{h-1}},$$
$$\mathcal{V} = vf^m + v_1\mathcal{F}_1^{m_1} + v_2\mathcal{F}_2^{m_2} + \cdots + v_{h-1}\mathcal{F}_{h-1}{}^{m_{h-1}},$$

where u, $u_1, \ldots, u_h$ and v, $v_1, \ldots, v_h$ are indeterminate parameters. The quantities $\mathcal{U}$, $\mathcal{V}$ are objects related to the covariants, but are not real covariants because the weights of the individual summands are different. Their resultant is a certain function of the $a_0, \ldots, a_n$ and the u and v; let it be

$$\mathcal{R}(\mathcal{U}, \mathcal{V}) = \sum \mathcal{I}_\rho \mathcal{P}_\rho(u, v),$$

where the $\mathcal{P}_\rho$ are products of the u, v. Then the $\mathcal{I}_\rho$ are invariants of the form f because $\mathcal{I}_\rho$ is an invariant of $\mathcal{U}$ and $\mathcal{V}$, hence an invariant of f, since now we have achieved homogeneity in the coefficients. This follows from the fact that $\mathcal{P}_\rho$ is a product of only the u and v, and the coefficients of u and v are always homogeneous in the a. But all $\mathcal{I}_\rho$ are equal to zero according to the hypothesis; hence $\mathcal{U}$ and $\mathcal{V}$ always have a common linear factor, no matter which values one assigns to the u, v. Therefore, all f, $\mathcal{F}_1$, $\mathcal{F}_2, \ldots, \mathcal{F}_{h(-1)}$ must contain the same linear factor, because otherwise one could always find values for u and v such that $\mathcal{U}$ and $\mathcal{V}$ would not contain a common linear factor, whereas $\mathcal{R}(\mathcal{U}, \mathcal{V}) = 0$. We transform the form f linearly in such a way that this factor becomes equal to x_2. Consequently, all sources of the covariants f, $\mathcal{F}_1, \ldots, \mathcal{F}_h$ have to be zero, and hence it follows successively that:

$$\begin{aligned}
&a_0 = 0, && \\
&a_1 = 0 && \text{because } a_0a_2 - a_1^2 = 0, \\
&a_2 = 0 && \text{because } a_0a_4 - 4a_1a_3 + 3a_2^2 = 0, \\
& && \cdots \\
&a_h = 0 && \text{because } a_0a_{2h} + \cdots \pm \frac{1}{2}\binom{2h}{h}a_h^2 = 0.
\end{aligned}$$

For odd n this follows from the previous line of reasoning; for even n one only has to observe in addition that the $2h$th transvection is an invariant, hence is equal to zero. Herewith, the first part of the theorem is proven.

The second part of the theorem is easily proven. We think of the form as being transformed in such a way that x_2 is the $(h+1)$-fold linear factor, so that

$$a_0 = 0,\ a_1 = 0,\ a_2 = 0, \ldots, a_h = 0.$$

An invariant has the form

$$\mathcal{I} = \sum Z a_0^{\nu_0} a_1^{\nu_1} \cdots a_n^{\nu_n},$$

and all terms containing a_0 or a_1 or ... or a_h vanish. For any eventual other terms we have

$$\begin{aligned} \nu_{h+1} + \nu_{h+2} + \cdots + \nu_n &= g, \\ (h+1)\nu_{h+1} + (h+2)\nu_{h+2} + \cdots + n\nu_n &= p \\ &= \frac{ng}{2}, \end{aligned}$$

so

$$2(h+1)\nu_{h+1} + 2(h+2)\nu_{h+2} + \cdots + 2n\nu_n = n\nu_{h+1} + n\nu_{h+2} + \cdots + n\nu_n.$$

The numerical coefficients of the ν on the left side are certainly bigger than those on the right side. Since all ν have to be ≥ 0, they all have to be equal to zero; so such terms cannot appear in $\mathcal{I}$ in the first place. Consequently, the invariant vanishes always—which proves the theorem.

This theorem and its proof indicate at the same time how to find a system of invariants of a binary base form, through which all others can be expressed integrally and algebraically. Such a system is formed by the invariants $\mathcal{I}_\rho$, respectively $\mathcal{I}_\rho$, $\mathcal{F}_h$, because, if these vanish, then the form has an $(h+1)$-fold linear factor, so that then—according to the second part of the theorem—all invariants vanish, allowing integral and algebraic representation of $\mathcal{I}_\rho$, respectively $\mathcal{I}_\rho$, $\mathcal{F}_h$. Thus, one only needs to construct the h transvections, the forms $\mathcal{U}$, $\mathcal{V}$, and their resultant. The coefficients of the products of the u, v then give—for odd n by themselves, for even n after adding $(f,f)_n$—the invariant system that allows integral algebraic representation of all invariants.

All the above observations can be generalized to simultaneous forms, and for the examples it is even advisable to use simultaneous forms rather than a single form, since the in- and covariants of a form of order higher than four are very complicated and numerous. However, the investigations along these lines have not been carried out yet for simultaneous forms, even though there do not seem to exist any fundamental difficulties. Only one case has often been considered, and settled, namely, the case of a simultaneous system of binary linear forms. Let

$$\begin{aligned} f_1 &= a_1x_1 + b_1x_2, \\ f_2 &= a_2x_1 + b_2x_2, \\ &\dots \\ f_n &= a_nx_1 + b_nx_2 \end{aligned}$$

be given n linear forms. Their full invariant system is given by the determinants

$$p_{ik} = a_ib_k - a_kb_i \quad (i, k = 1, 2, \dots, n).$$

But there is a large number of syzygies between them, since their number is much larger than $2n-3$. To obtain a system of independent invariants, we proceed as follows. We form the function

$$f_1t^{n-1} + f_2t^{n-2} + \cdots + f_n = x_1h(t) + x_2g(t),$$

which represents an involution determined by $h(t) = 0$, $g(t) = 0$. We form the functional determinant of h and g, that is,

$$\begin{aligned} (h, g)_1 = \mathcal{H}^{(2n-4)}(t) &= h'(t)g(t) - g'(t)h(t) = g(t)^2 \left[\frac{h}{g}\right]'(t) \\ &= A_0t^{2n-4} + A_1t^{2n-5} + \cdots + A_{2n-4}. \end{aligned}$$

The coefficients A, being linear combinations of the p_{ik}, are themselves invariants of the base forms. If these invariants A vanish, then either all coefficients of h are zero, or those of g, or h and g are equal up to a numerical factor. In all these cases, all p_{ik} are equal to zero, and hence $A_0, A_1, \dots, A_{2n-4}$ form a system of invariants through which all others can be expressed integrally and algebraically. At the same time, these are independent of each other, since there are $2n-3$ of them. So, it is not difficult, for a simultaneous system of linear forms, to find a full system of invariants for integral and algebraic representation.

Lecture XLV (July 27, 1897)

II.5 The ternary nullform

The concepts and observations we have been dealing with so far can be extended in two directions. First, instead of considering binary forms in one sequence of variables, one can consider those in two sequences of variables, that is, consider forms of the type

$$f(x_1, x_2;\ \xi_1, \xi_2),$$

such that f is homogeneous of order n in the x, and homogeneous of order ν in the ξ. One can then specify different transformations for the ξ; for instance,

$$\begin{aligned} x_1 &= \alpha x_1' + \beta x_2', \\ x_2 &= \gamma x_1' + \delta x_2'; \end{aligned} \qquad (\epsilon = \alpha\delta - \beta\gamma)$$

$$\begin{aligned} \xi_1 &= \rho \xi_1' + \sigma \xi_2', \\ \xi_2 &= \tau \xi_1' + \lambda \xi_2', \end{aligned} \qquad (\kappa = \rho\lambda - \sigma\tau)$$

where the transformation coefficients are all independent of each other. Invariants of f are then those integral rational functions of the coefficients which, up to a product of powers of ϵ and κ, remain unchanged if one replaces the original coefficients by the transformed ones. The latter ones have been investigated, in particular when $\nu = 1$, which leads one to the so-called *combinants*. Generally, one can ask the same questions as before, for instance concerning the nullforms, which would be especially important. Furthermore, one can introduce more sequences of variables, and ask the same questions as before. The easiest case to consider here would be the one in which all orders are equal to one. Our general theorems remain completely valid in this case. It only would remain to be seen which modifications need to be introduced.

The second extension consists of considering more than two variables. One can also go further and consider forms in arbitrarily many variable sequences, where each such sequence contains an arbitrary number of variables, and each sequence is subjected to a different general transformation. For instance, in the form

$$f(x_1, x_2, x_3;\ \xi_1, \xi_2),$$

where the order in the x is n and the order in the ξ is ν, the x (respectively ξ) can be subjected to the independent transformations

$$\begin{aligned} x_1 &= \alpha_{11} x_1' + \alpha_{12} x_2' + \alpha_{13} x_3', \\ x_2 &= \alpha_{21} x_1' + \alpha_{22} x_2' + \alpha_{23} x_3', \\ x_3 &= \alpha_{31} x_1' + \alpha_{32} x_2' + \alpha_{33} x_3'; \\ \xi_1 &= \beta_{11} \xi_1' + \beta_{12} \xi_2', \\ \xi_2 &= \beta_{21} \xi_1' + \beta_{22} \xi_2', \end{aligned}$$

and one can make entirely analogous observations as before. The general theorems about invariant fields hold true here as well. On the other hand, as soon as one leaves binary forms, the construction of the nullforms is not possible any more in the same way as in the previous section for binary forms. But here too one can find a general method which can immediately be applied to the most general case, as soon as one understands it for a ternary form. We want to discuss this method and then briefly indicate its proof. We consider the ternary form

$$f(x_1, x_2, x_3) = \sum a_{n_1 n_2 n_3} x_1^{n_1} x_2^{n_2} x_3^{n_3},$$

with $n_1 + n_2 + n_3 = n$, and subject it to the linear transformation

$$\begin{aligned} x_1 &= \alpha_{11} x_1' + \alpha_{12} x_2' + \alpha_{13} x_3', \\ x_2 &= \alpha_{21} x_1' + \alpha_{22} x_2' + \alpha_{23} x_3', \\ x_3 &= \alpha_{31} x_1' + \alpha_{32} x_2' + \alpha_{33} x_3', \end{aligned}$$

so that it becomes

$$f'\left(x_1', x_2', x_3'\right) = \sum a'_{n_1 n_2 n_3} {x_1'}^{n_1} {x_2'}^{n_2} {x_3'}^{n_3},$$

with $n_1 + n_2 + n_3 = n$. If we set

$$\delta = \begin{vmatrix} \alpha_{11} & \alpha_{12} & \alpha_{13} \\ \alpha_{21} & \alpha_{22} & \alpha_{23} \\ \alpha_{31} & \alpha_{32} & \alpha_{33} \end{vmatrix},$$

then an invariant of f is defined by the equation:

$$\mathcal{I}(a') = \delta^p \mathcal{I}(a).$$

It has the analogous elementary properties that hold for binary forms and their invariants.

The nullforms can then be established in the following way. Each null form, through a suitable linear transformation, can be brought into such a form that certain coefficients vanish, while the others remain arbitrary and independent of each other. Conversely, every such form, in which certain prescribed coefficients are zero, and the others are arbitrary, is a nullform. Thus—just as for binary forms—one can find a number of *canonical nullforms*, which are null forms for any values of the arbitrary coefficients, and such that, conversely, every nullform can be linearly transformed into one of these canonical forms.

The canonical nullforms can be found as follows. Take an equilateral triangle of side length n (see Fig. 1, for which $n = 5$), and divide each

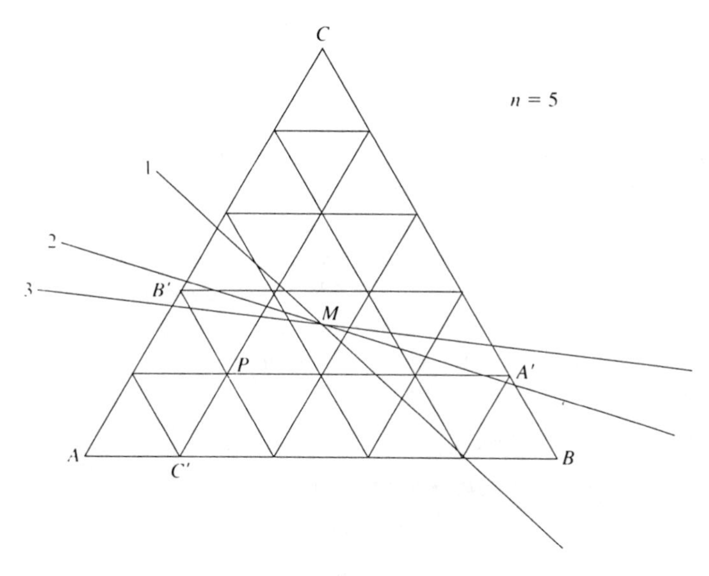

Fig. 1.

side into n equal parts, of which each part has length one. Draw the parallels to the other sides through each dividing point, so that the whole triangle is partitioned into equilateral triangles of side length one. For an arbitrary vertex P of such a small triangle, that is, for each so-called lattice point, one can choose as coordinates the lengths of the sides $\mathcal{P}A'$, $\mathcal{P}B'$, $\mathcal{P}C'$, where A', B', C', respectively, are determined on the sides BC, CA, AB of the whole triangle as intersection points of the parallels to these sides which pass through $\mathcal{P}$.

Each point corresponds uniquely to a coordinate triple, and to each coordinate triple corresponds uniquely a lattice point, and we remark that these coordinates are always positive integers $\leq n$, whose sum is n. Therefore, to each point of the lattice there corresponds, in a one-to-one fashion, a term of the ternary form of order n

$$f = \sum a_{n_1 n_2 n_3} x_1^{n_1} x_2^{n_2} x_3^{n_3}.$$

Namely, the term $a_{n_1 n_2 n_3} x_1^{n_1} x_2^{n_2} x_3^{n_3}$ shall correspond to the lattice point with coordinates n_1, n_2, n_3.

Now, if M is the center of the whole equilateral triangle, then we obtain precisely all canonical nullforms according to the following rule.

Rule *Draw an arbitrary straight line through the center M, and then determine all those vertices which lie on this line or on the side of A*

off the line. The coefficients $a_{n_1n_2n_3}$ in the ternary form of order n, corresponding to these vertices n_1, n_2, n_3, are to be set equal to zero, while the other coefficients are left arbitrary.

Here, A is an arbitrary but fixed vertex. Since in this way one obtains all canonical nullforms, it follows that the number of different kinds of nullforms is equal to the number of essentially different positions that the line through M can assume with respect to the vertices under consideration. Herein, one need not consider those positions for which the corresponding forms are special canonical nullforms.

For $n = 2$ up to $n = 5$ we obtain the following types of canonical nullforms:

$$\begin{array}{lll}
n=2: & 1. & (x_2x_3)_2, \\
 & 2. & ax_2x_1 + x_2(x_2x_3)_1; \\
n=3: & & ax_3^2x_1 + (x_2x_3)_3; \\
n=4: & 1. & (x_2x_3)_3x_1 + (x_2x_3)_4, \\
 & 2. & x_2\{ax_2x_1^2 + x_2x_1(x_2x_3)_1 + (x_2x_3)_3\}; \\
n=5: & 1. & ax_2^3x_1^2 + x_1(x_2x_3)_4 + (x_2x_3)_5, \\
 & 2. & x_2\{x_2x_1^2(x_2x_3)_1 + x_2x_1(x_2x_3)_2 + (x_2x_3)_4\}, \\
 & 3. & x_2^2\{ax_1^3 + x_1^2(x_2x_3)_1 + (x_2x_3)_2x_1 + (x_2x_3)_3\}.
\end{array}$$

Here, the expression $(x_2x_3)_s$ denotes a general expression of order s in x_2, x_3 with arbitrary coefficients (a binary form). It needs to be remarked that, for $n = 2$, the two canonical nullforms can be transformed into each other through a linear transformation, so in this case there really is only one nullform. The figures for $n = 2$ up to $n = 4$ correspond to Figs. 2, 3, and 4, respectively. For $n = 5$, we obtain Fig. 1.

Therefore, one can also say easily what the vanishing of all invariants means geometrically for the curve $f = 0$. If $n = 2$, the curve breaks up into two lines; if $n = 3$, the curve has a cusp. Further, if $n = 4$, then all invariants vanish if and only if the curve $f = 0$ either has a point of multiplicity three or else decomposes into a cubic curve and an inflectional tangent of it.

Lecture XLVI (July 29, 1897)

This very simple and clear method to determine the nullforms can also be expanded to forms with more variables. For quaternary forms, for

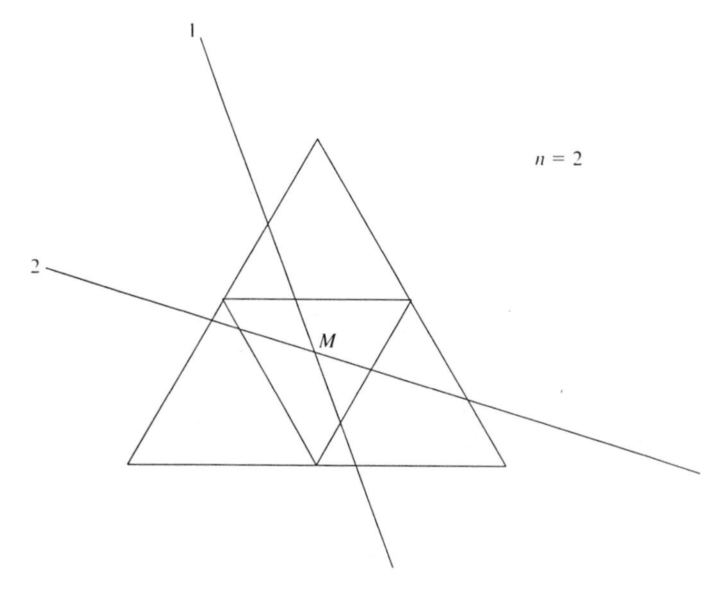

Fig. 2.

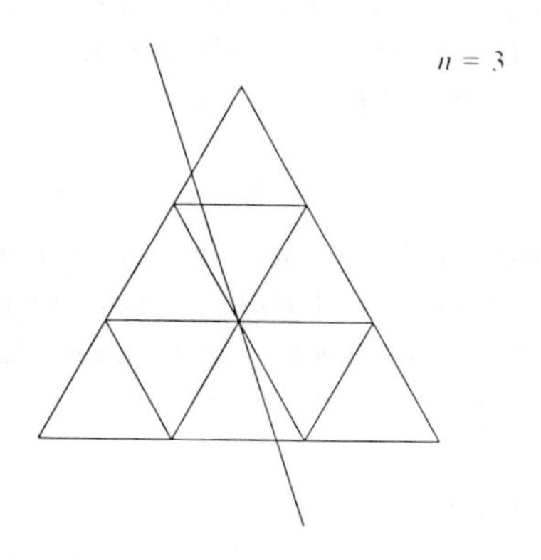

Fig. 3.

instance, one has to consider a regular tetrahedron of edge length n, has to divide each edge into n equal parts, and then has to divide the tetrahedron with parallel planes through these division points into small regular tetrahedra of edge length one. Then put a plane through the center of the tetrahedron. The rest is the same as before. If $n = 3$, for instance, one obtains the following:

Theorem *The invariants of a quaternary cubic form f vanish if and only if the surface defined by $f = 0$ has a point of multiplicity two for*

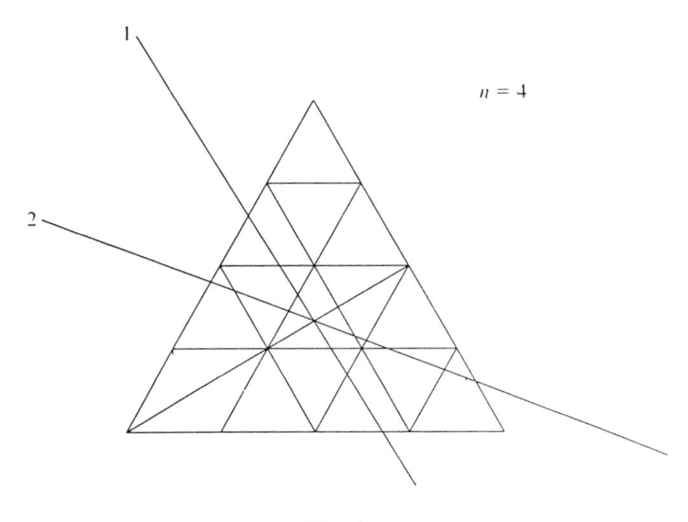

Fig. 4.

which the quadratic polar is a plane which is counted twice; or if it has a point of multiplicity two, for which the quadratic polar consists of two distinct planes whose line of intersection is completely contained in the surface.

Regarding the proof of this theorem for the ternary form, we give an indication of the principles on which it is based. As can be shown easily (see Hilbert 1893), the earlier geometric representation of the canonical nullform has the following analytic interpretation.

Theorem *The form is a canonical nullform if and only if one can find three integers λ_1, λ_2, λ_3 whose sum $\lambda_1 + \lambda_2 + \lambda_3$ is negative, such that each coefficient $a_{n_1 n_2 n_3}$ becomes zero if $\lambda_1 n_1 + \lambda_2 n_2 + \lambda_3 n_3 < 0$ is negative, while the other coefficients remain arbitrary.**

It is easily seen that the stated condition is sufficient. Namely, suppose we are given an invariant

$$\mathcal{I} = \sum Z \prod_{n_1 n_2 n_3} a_{n_1 n_2 n_3}^{e_{n_1 n_2 n_3}}$$

* In the general setting of geometric invariant theory, this theorem is called the Hilbert–Mumford Numerical Criterion.

$\mathcal{I}$ is a homogeneous isobaric function, and one has—analogously to binary forms—for each term of the sum:

$$\sum_{n_1 n_2 n_3} n_1 e_{n_1 n_2 n_3} = \frac{ng}{3},$$

$$\sum_{n_1 n_2 n_3} n_2 e_{n_1 n_2 n_3} = \frac{ng}{3},$$

$$\sum_{n_1 n_2 n_3} n_3 e_{n_1 n_2 n_3} = \frac{ng}{3},$$

where g is the degree of the invariant. Multiplying the three equations by λ_1, λ_2, λ_3, respectively, and adding the resulting equations gives

$$\sum_{n_1 n_2 n_3} e_{n_1 n_2 n_3} \left(\lambda_1 n_1 + \lambda_2 n_2 + \lambda_3 n_3\right) = \frac{ng}{3} \left(\lambda_1 + \lambda_2 + \lambda_3\right),$$

or, if we set

$$\lambda_1 + \lambda_2 + \lambda_3 = -\lambda,$$

where λ denotes a positive integer, then

$$\sum_{n_1 n_2 n_3} e_{n_1 n_2 n_3} \left(\lambda_1 n_1 + \lambda_1 n_2 + \lambda_3 n_3\right) = -\frac{ng}{3}\lambda.$$

If we omit the terms on the left for which $\lambda_1 n_1 + \lambda_1 n_2 + \lambda_3 n_3$ is positive or zero, then one has

$$\sum_{n_1 n_2 n_3}{}' e_{n_1 n_2 n_3} \left(\lambda_1 n_1 + \lambda_2 n_2 + \lambda_3 n_3\right) \leq -\frac{ng}{3}\lambda.$$

The sum $\sum'$ extends only over those systems n_1, n_2, n_3 for which $\lambda_1 n_1 + \lambda_2 n_2 + \lambda_3 n_3$ is negative. Therefore, we also have

$$\sum_{n_1 n_2 n_3}{}' e_{n_1 n_2 n_3} \left|\lambda_1 n_1 + \lambda_2 n_2 + \lambda_3 n_3\right| \geq \frac{ng}{3}\lambda.$$

On the left is an integer, on the right a number > 0, so it follows that

$$\sum_{n_1 n_2 n_3}{}' e_{n_1 n_2 n_3} \left|\lambda_1 n_1 + \lambda_2 n_2 + \lambda_3 n_3\right| \geq 1.$$

So we have that in each term of the invariant there is at least one of the coefficients $a_{n_1 n_2 n_3}$ of $\mathcal{I}$ which are set equal to zero, that occurs to a positive power. Because, if all powers $e_{n_1 n_2 n_3}$ would be zero, then $\sum' = 0$, which cannot be. Thus, the invariant is also equal to zero, which proves the first half of the theorem.

The converse is substantially more difficult to prove. We only remark here that one uses the following:

Fundamental Theorem *A given ternary base form with certain* **numerical** *coefficients has a nonzero invariant if and only if the substitution determinant*

$$\delta = \begin{vmatrix} \alpha_{11} & \alpha_{12} & \alpha_{13} \\ \alpha_{21} & \alpha_{22} & \alpha_{23} \\ \alpha_{31} & \alpha_{32} & \alpha_{33} \end{vmatrix}$$

is an integral algebraic function of the transformed coefficients.

We shall prove this theorem. Suppose that the ternary form

$$f(x_1, x_2, x_3) = a_1 x_1^n + \cdots + a_N x_3^n$$

is carried into

$$f'(x_1', x_2', x_3') = a_1' {x_1'}^n + \cdots + a_N' {x_3'}^n$$

by the transformation

$$\begin{aligned} x_1 &= \alpha_{11} x_1' + \alpha_{12} x_2' + \alpha_{13} x_3', \\ x_2 &= \alpha_{21} x_1' + \alpha_{22} x_2' + \alpha_{23} x_3', \\ x_3 &= \alpha_{31} x_1' + \alpha_{32} x_2' + \alpha_{33} x_3'. \end{aligned}$$

We think of the a as numerical quantities; then the a' are forms of degree n in the α, with numerical coefficients. If f has a nonzero invariant, then δ is an integral algebraic function of the a'. That is evident. Namely, if that invariant is $\mathcal{I}$, then we have

$$\delta^p \mathcal{I}(a_1, \ldots, a_N) = \mathcal{I}(a_1', \ldots, a_N'),$$

where $p \neq 0$, since there does not exist an invariant of weight zero. Thus, indeed, one has:

$$\delta^p - \frac{\mathcal{I}(a_1', \ldots, a_N')}{\mathcal{I}(a_1, \ldots, a_N)} = 0.$$

Conversely, if δ satisfies an equation of the form

$$\delta^m + \mathcal{G}_1(a_1', \ldots, a_N')\,\delta^{m-1} + \cdots + \mathcal{G}_m(a_1', \ldots, a_N') = 0,$$

where the $\mathcal{G}$ are integral rational functions of the a', then we certainly also have

$$\Omega^m \delta^m + \Omega^m \left\{\mathcal{G}(a_1', \ldots, a_N')\,\delta^{m-1}\right\} + \cdots + \Omega^m \mathcal{G}_m(a_1', \ldots, a_N') = 0.$$

Here we set

$$\Omega = \begin{vmatrix} \frac{\partial}{\partial \alpha_{11}} & \frac{\partial}{\partial \alpha_{12}} & \frac{\partial}{\partial \alpha_{13}} \\ \frac{\partial}{\partial \alpha_{21}} & \frac{\partial}{\partial \alpha_{22}} & \frac{\partial}{\partial \alpha_{23}} \\ \frac{\partial}{\partial \alpha_{31}} & \frac{\partial}{\partial \alpha_{32}} & \frac{\partial}{\partial \alpha_{33}} \end{vmatrix} = \frac{\partial^3}{\partial \alpha_{11} \partial \alpha_{22} \partial \alpha_{33}} + \cdots .$$

The laws which we derived for this symbol in the binary case (Section II.2) remain valid here too without change. Therefore, one has

$$\Omega^m \delta^m + \mathcal{I}_1 (a_1, \ldots, a_N) + \mathcal{I}_2 (a_1, \ldots, a_N) + \cdots + \mathcal{I}_m (a_1, \ldots, a_N) = 0,$$

where the $\mathcal{I}$ are invariants. Since $\Omega^m \delta^m \neq 0$, not all $\mathcal{I}$ can vanish simultaneously—which was to be proven.

With the help of this theorem we can now carry out the proof of the converse, stated above; however, it is still very cumbersome and difficult. In the process, one has to use theorems about substitutions, whose coefficients are power series, and these theorems are difficult to derive in themselves. Thus, we can only continue to refer to Hilbert (1893).

We should make one more remark. Every nullform can be transformed linearly into a canonical nullform. If we call a "class of forms" the totality of all those forms which can be transformed into each other through a linear transformation with nonvanishing determinants, then one has the following:

Theorem *Each class of nullforms contains a canonical nullform.*

Lecture XLVII (July 30, 1897)

II.6 The finiteness of the number of irreducible syzygies and of the syzygy chain

With regard to the fundamental questions, we need to take a closer look at the syzygies. We again fix a binary form

$$f(x_1, x_2) = a_0 x_1^n + \cdots + a_n x_2^n,$$

even though the observations are still valid in general. The full invariant system is finite, as we have proven; that means one can choose from among the invariants a finite number i_1, $i_2, \ldots, i_N$ so that all others are integral rational functions of these. At the same time these N invariants are still dependent on each other, in that—at least in general—there exist a number of syzygies among them, that is, equations of the type:

$$X_1 = 0,\ X_2 = 0,\ X_3 = 0, \dots,$$

where X_1, X_2, $X_3, \dots$ are integral rational functions of i_1, $i_2, \dots, i_N$, and these equations hold identically in the coefficients a_i of the base form. The number of syzygies is inherently always infinite—if there is *one* at all—because from $X_1 = 0$ it follows that

$$A_1 X_1 = 0,$$

where A_1 is an arbitrary function of i_1, $i_2, \dots, i_N$, and this equation holds identically. In general,

$$A_1 X_1 + A_2 X_2 + A_3 X_3 + \cdots = 0$$

is a new syzygy, when $A_1, A_2, A_3, \dots$ are integral rational functions of $i_1, i_2, \dots, i_N$. By an irreducible syzygy we mean one whose left side—the so-called syzygant—cannot be obtained as a linear combination of syzygants of lower degrees. We then have the following:

Theorem *A finite system of invariants possesses only a finite number of irreducible syzygies. In particular, between the invariants of a system of base forms there always exists only a finite number of irreducible syzygies.*

This theorem is obvious from the Fundamental Theorem in Section II.2. Indeed, one thinks of all syzygies as ordered according to the degree of the invariants—the number of syzygies of a given degree must be finite because there are only finitely many invariants. Then one can find a number m such that each syzygant X can be brought into the form:

$$X = A_1 X_1 + A_2 X_2 + A_3 X_3 + \cdots + A_m X_m,$$

where $X_1, X_2, \dots, X_m$ are the first m syzygants. The theorem was only stated for forms, but is valid for arbitrary integral rational functions, which can be seen by setting one variable equal to one.

It is, furthermore, obvious that one can always choose $A_1, A_2, \dots, A_m$ so that the relation $X = 0$ holds identically not only in the coefficients a, but even in the invariants $i_1, i_2, \dots, i_N$, considered as new variables. This must always occur. Such a syzygy is called a *second-order syzygy.* For instance, one only needs to set $A_2 = X_1$ and $A_1 = -X_2$ to obtain an example. In some sense these syzygies are, therefore, not syzygies any more.

To determine the second-order syzygies, one therefore needs to determine the coefficients A in such a way that the equation $X = 0$ becomes an identity in the $i_1, i_2, \ldots, i_N$. Suppose that

$$\begin{aligned} A_1 &= X_{1j}, \\ A_2 &= X_{2j}, \\ &\cdots \\ A_m &= X_{mj} \end{aligned}$$

is a general system of solutions. Then, again, the number of irreducible solution systems is finite. That is, the number of irreducible second-order syzygies is finite; or: It is possible to choose m' solution systems such that any other one can be expressed linearly in terms of these, and can hence be brought into the form

$$\begin{aligned} A_1 &= B_1X_{11} + B_2X_{12} + \cdots + B_{m'}X_{1m'}, \\ A_2 &= B_1X_{21} + B_2X_{22} + \cdots + B_{m'}X_{2m'}, \\ &\cdots \\ A_m &= B_1X_{m1} + B_2X_{m2} + \cdots + B_{m'}X_{mm'}, \end{aligned}$$

where the B are again arbitrary functions of the $i_1, i_2, \ldots, i_N$. Obviously, one can proceed further in this fashion and ask about the solutions of the equations

$$B_1X_{j1} + B_2X_{j2} + \cdots + B_{m'}X_{jm'} = 0, \qquad j = 1, \ldots, m,$$

for the B, so that these equations hold identically in the $i_1, i_2, \ldots, i_N$. In this way, one obtains the "third-order syzygies." Their number is finite if one does not count the solutions that can be expressed linearly in terms of the others. One can continue in this way and obtain fourth-, fifth-, … order syzygies. Now, however, we have two important theorems. First, one has the following:

Theorem *The chain of syzygies terminates after a finite number of steps.*

If we assume that there still are syzygies of the ρth kind, but no more of $(\rho+1)$th kind or higher, then we always have that $\rho \leq N+1$, where N is the number of invariants of the full invariant system. Incidentally, this theorem is very difficult to prove; we cannot go into the proof (see Hilbert 1890).*

* This is the Hilbert Syzygy Theorem. An elementary proof using Gröbner bases can be found in D. Eisenbud, *Commutative Algebra with a View Toward Algebraic Geometry*, to appear.

It is essential to know the number of syzygies of a given order when one asks about the number of invariants of a given degree R; let this number be $\chi(R)$, where we are considering the invariants between which there are no more relations. Further thought shows that one obtains the number $\chi(R)$ in the following way. Let $i_1, i_2, \ldots, i_N$ be the full invariant system. Then determine all integral rational functions of these invariants which have degree R in the a_i in such a way that there do not exist any linear relations with constant coefficients between them. From this one needs to subtract the number of independent first-order syzygies, add the second-order ones, then subtract the third-order syzygies, etc.—a calculation which has to terminate, according to the above theorem. One then finds the important theorem:

Theorem *The number $\chi(R)$ of linearly independent invariants of degree R is an integral rational function of R; this function is called the* **characteristic function.***

Incidentally, these observations do not depend on the fact that one is dealing with invariants; they can be made whenever one is concerned with the algebraic connection between given quantities. We explain this in terms of the following example, which should be worked out completely. We proceed from the following general theorem, which is an immediate consequence of the general theorem in Section II.2 (see loc. cit. Hilbert (1890)).

Theorem *Through a given algebraic space curve one can place a finite number m of surfaces*

$$\mathcal{F}_1 = 0,\ \mathcal{F}_2 = 0, \ldots, \mathcal{F}_m = 0$$

in such a way that every other surface that contains the space curve can be represented by an equation of the form

$$A_1\mathcal{F}_1 + A_2\mathcal{F}_2 + \cdots + A_m\mathcal{F}_m = 0,$$

where $A_1, A_2, \ldots, A_m$ denote quaternary forms.

* The function $\chi(R)$ is the Hilbert function of the invariant ring. The preceding paragraph gives its representation as the alternating sum of the Hilbert functions of the free modules in the minimal free resolution.

Suppose given the cubic space curve

$$\begin{aligned} x_1 &= t^3, \\ x_2 &= t^2, \\ x_3 &= t, \\ x_4 &= 1. \end{aligned}$$

Through this curve pass the three surfaces

$$\mathcal{F}_1 = 0,\ \mathcal{F}_2 = 0,\ \mathcal{F}_3 = 0,$$

where

$$\begin{aligned} \mathcal{F}_1 &\equiv x_1 x_3 - x_2^2, \\ \mathcal{F}_2 &\equiv x_2 x_3 - x_1 x_4, \\ \mathcal{F}_3 &\equiv x_2 x_4 - x_3^2. \end{aligned}$$

Every other surface that contains the curve can be represented in the form

$$A_1 \mathcal{F}_1 + A_2 \mathcal{F}_2 + A_3 \mathcal{F}_3 = 0,$$

where the A are forms of the same degree in x_1, x_2, x_3, x_4. To prove this, let

$$\mathcal{F} = \sum C x_1^{r_1} x_2^{r_2} x_3^{r_3} x_4^{r_4}$$

be a form that becomes identically zero under the substitution $x_1 = t^3, x_2 = t^2, x_3 = t, x_4 = 1$. We set

$$\Phi \equiv \Psi \mod (\mathcal{F}_1, \mathcal{F}_2, \mathcal{F}_3)$$

if

$$\Phi - \Psi = A_1 \mathcal{F}_1 + A_2 \mathcal{F}_2 + A_3 \mathcal{F}_3.$$

Then, obviously,

$$\begin{aligned} x_1 x_3 &\equiv x_2^2 \mod (\mathcal{F}_1, \mathcal{F}_2, \mathcal{F}_3), \\ x_1 x_4 &\equiv x_2 x_3 \mod (\mathcal{F}_1, \mathcal{F}_2, \mathcal{F}_3), \\ x_2 x_4 &\equiv x_3^2 \mod (\mathcal{F}_1, \mathcal{F}_2, \mathcal{F}_3), \end{aligned}$$

and with the help of these congruences, $\mathcal{F}$ can be reduced to

$$\mathcal{F} \equiv \sum C x_1^{\kappa_1} x_2^{\kappa_2} + \sum C' x_2^{\lambda_2} x_3^{\lambda_3} + \sum C'' x_3^{\mu_3} x_4^{\mu_4} \mod (\mathcal{F}_1, \mathcal{F}_2, \mathcal{F}_3).$$

This is because, using $x_1 x_3 \equiv x_2^2$, one can remove either x_1 or x_3 from $x_1^{r_1} x_2^{r_2} x_3^{r_3} x_4^{r_4}$. In the first case one keeps $x_2^{()} x_3^{()} x_4^{()}$; but then one can remove either x_2 or x_4 with the help of the congruence $x_2 x_4 \equiv x_3^2$, and one keeps either $x_3^{(\,)}x_4^{(\,)}$ or $x_2^{(\,)}x_3^{(\,)}$. In the other case, one has $x_1^{(\,)}x_2^{(\,)}x_4^{(\,)}$, and because $x_1x_4 \equiv x_2x_3$ one arrives either at $x_2^{(\,)}x_3^{(\,)}x_4^{(\,)}$—that is, at the previous case—or at $x_1^{(\,)}x_2^{(\,)}x_3^{(\,)}$, which can be reduced to $x_1^{(\,)}x_2^{(\,)}$ or $x_2^{(\,)}x_3^{(\,)}$ via $x_1x_3 \equiv x_2^2$. Because of homogeneity one now has

$$\kappa_1 + \kappa_2 = \lambda_2 + \lambda_3 = \mu_3 + \mu_4,$$

hence also

$$3\kappa_1 + 2\kappa_2 > 2\lambda_2 + \lambda_3 > \mu_3.$$

If one now introduces t into the form $\mathcal{F}$, then it follows that

$$\mathcal{F}(t^3, t^2, t, 1) = \sum C t^{3\kappa_1 + 2\kappa_2} + \sum C' t^{2\lambda_2 + \lambda_3} + \sum C'' t^{\mu_3},$$

which is supposed to be identically equal to 0. But, because of the earlier inequalities, no term of one sum can pair up with a term of another sum; hence all C, C', C'' have to be equal to zero—which proves the assertion.

Lecture XLVIII (August 2, 1897)

We now want to further concern ourselves with the question of how many surfaces of a given order R pass through the space curve so that there are no linear relations between them. We first have to establish the independent systems of equations that correspond to the higher order syzygies. Among the forms

$$\mathcal{F} = A_1\mathcal{F}_1 + A_2\mathcal{F}_2 + A_3\mathcal{F}_3$$

there are those that are not essentially different from each other and these are given by the equation

$$A_1\mathcal{F}_1 + A_2\mathcal{F}_2 + A_3\mathcal{F}_3 = B_1\mathcal{F}_1 + B_2\mathcal{F}_2 + B_3\mathcal{F}_3$$

or, if one sets $A_1 - B_1 = X_1, \ldots,$ by

$$X_1\mathcal{F}_1 + X_2\mathcal{F}_2 + X_3\mathcal{F}_3 = 0.$$

This equation is certainly satisfied by

$$X_1 = x_3, \quad X_2 = x_2, \quad X_3 = x_1; \qquad \text{and by} \qquad X_1 = x_4, \quad X_2 = x_3, \quad X_3 = x_2.$$

These are the only solutions; any others can be expressed in the form

$$X_1 = x_3\mathcal{Y}_1 + x_4\mathcal{Y}_2,$$
$$X_2 = x_2\mathcal{Y}_1 + x_3\mathcal{Y}_2,$$
$$X_3 = x_1\mathcal{Y}_1 + x_2\mathcal{Y}_2,$$

where the $\mathcal{Y}$ are arbitrary forms of the same order. This is because every solution X_1 can obviously be written in the form

$$X_1 = C_1 + C_2 x_3,$$

where C_1 does not contain x_3 any more. But we obtain a new solution by adding $-x_3C_2$, $-x_2C_2$, or $-x_1C_2$, respectively, to this solution. The solution $X_1 = C_1 \ldots$ obtained in this way is such that X_1 does not contain x_3 any more. Similarly, we can also eliminate x_4 from X_1. Thus, each solution can be reduced in such a way that X_1 contains neither x_3 nor x_4. The equation

$$X_1\mathcal{F}_1 + X_2\mathcal{F}_2 + X_3\mathcal{F}_3 = 0$$

is supposed to hold identically in the x. Hence, if we set $x_3 = 0$, $x_4 = 0$, then $\mathcal{F}_2 = 0$, $\mathcal{F}_3 = 0$, $\mathcal{F}_1 = -x_2^2 \neq 0$; so we must have

$$X_1 \equiv 0$$

identically. Thus, every solution can be reduced to a solution with $X_1 = 0$, modulo the two special solutions. The remaining equation is

$$X_2\mathcal{F}_2 + X_3\mathcal{F}_3 = 0$$

whose general solution is

$$X_2 = \mathcal{Y}\mathcal{F}_3 = \mathcal{Y}\left(x_2x_4 - x_3^2\right),$$
$$X_3 = \mathcal{Y}(-\mathcal{F}_2) = \mathcal{Y}\left(x_1x_4 - x_2x_3\right),$$

which is clearly representable in the form

$$X_2 = x_4\mathcal{Y} \cdot x_2 - x_3\mathcal{Y} \cdot x_3,$$
$$X_3 = x_4\mathcal{Y} \cdot x_1 - x_3\mathcal{Y} \cdot x_2,$$

where

$$X_1 = x_4\mathcal{Y} \cdot x_3 - x_3\mathcal{Y} \cdot x_4$$

is indeed equal to zero. Thus, every solution is a linear combination of those two.

We must further investigate how often such a solution system occurs for different values of $\mathcal{Y}_1$ and $\mathcal{Y}_2$. So we must solve the following system of equations for the $\mathcal{Y}$:

$$\begin{aligned} x_3\mathcal{Y}_1 + x_4\mathcal{Y}_2 &= 0, \\ x_2\mathcal{Y}_1 + x_3\mathcal{Y}_2 &= 0, \\ x_1\mathcal{Y}_1 + x_2\mathcal{Y}_2 &= 0. \end{aligned}$$

From this follows

$$\mathcal{Y}_1 = -x_4 Z,\ \mathcal{Y}_2 = x_3 Z$$

and further

$$\left(-x_2x_4 + x_3^2\right) Z = 0,$$

hence $Z = 0$. That is, the system of equations does not have any solutions any more, so the chain of syzygies terminates herewith.

We now proceed to the actual determination of the number of linearly independent surfaces of order R that contain the space curve. The number of coefficients of a quaternary form of this order is (compare Section I.1) equal to $\frac{1}{6}(R+1)(R+2)(R+3)$. To determine the number $\chi(R)$ of conditions that a surface has to satisfy in order to pass through the space curve, one has to subtract from this number the number of linearly independent forms $\mathcal{F}$ of order R that can be represented by the formula

$$\mathcal{F} = A_1\mathcal{F}_1 + A_2\mathcal{F}_2 + A_3\mathcal{F}_3.$$

The latter number we obtain in turn by diminishing the total number of terms in the three forms A_1, A_2, A_3 of order $R-2$, that is, the number $3\cdot\frac{1}{6}(R-1)R(R+1)$, by the number of those linearly independent form systems X_1, X_2, X_3 of order $R-2$ that satisfy the equation

$$X_1\mathcal{F}_1 + X_2\mathcal{F}_2 + X_3\mathcal{F}_3 = 0.$$

This number is equal to the number of terms in $\mathcal{Y}_1, \mathcal{Y}_2$ which have order $R-3$; this number of terms, therefore, is $2\cdot\frac{1}{6}(R-2)(R-1)R$. According to the previous discussion, nothing further needs to be subtracted. The desired number is, therefore,

$$\begin{aligned} \chi(R) &= \tfrac{1}{6}(R+1)(R+2)(R+3) - 3\cdot\tfrac{1}{6}(R-1)R(R+1) \\ &\quad + 2\cdot\tfrac{1}{6}(R-2)(R-1)R \\ &= 1 + 3R. \end{aligned}$$

All the constants appearing in this formula are of great importance. Namely, the degree of the characteristic function gives in general the dimension of the object under consideration, the coefficient of the highest power gives the order of the object, and the other coefficients are in a close relationship to the genus of the object. For curves, the second coefficient directly gives the genus raised by one. Thus, our curve is of genus zero. The last remark, however, is only valid for curves without double points, etc.*

Lecture IL (August 3, 1897)

As we already remarked, this kind of procedure can be applied quite generally when it comes to algebraic relations between arbitrary quantities. We consider a further geometric example. For a triangle there are altogether nine items to consider, namely, the sides a, b, c; the cosines of the angles $\cos\alpha = c_\alpha, \ldots$; and the sines of the angles $\sin\alpha = s_\alpha, \ldots$.

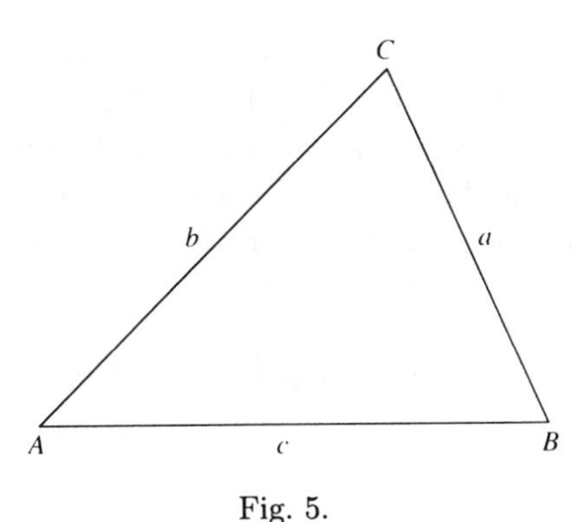

Fig. 5.

Between these nine items

$$a, b, c, c_\alpha, c_\beta, c_\gamma, s_\alpha, s_\beta, s_\gamma$$

there exist lots of algebraic relations; but, according to the cited theorem, there must be a finite number of linearly independent relations. Such syzygies are, for instance,

$$c_\alpha^2 + s_\alpha^2 - 1 = 0, \quad c_\beta^2 + s_\beta^2 - 1 = 0, \ldots;$$
$$as_\beta - bs_\alpha = 0, \quad bs_\gamma - cs_\beta = 0, \ldots;$$
$$c^2 - a^2 - b^2 + 2abc_\gamma = 0, \ldots.$$

* See Chapter I.7, in particular Ex. 7.2, of Hartshorne, *Algebraic Geometry*, Graduate Texts in Mathematics 52, Springer-Verlag, New York, 1977.

Now, let $S_1 = 0, S_2 = 0, \ldots, S_r = 0$ be a complete system of such syzygies. Then one can ask further about the second-order syzygies, that is, after syzygies of the type:

$$A_1 S_1 + A_2 S_2 + \cdots + A_r S_r = 0,$$

which hold identically in those nine quantities. In essence, there is again only a finite number. Likewise, one is led further to third-order syzygies, etc. This sequence terminates at some point. Only when one knows all independent higher order syzygies does one completely understand the algebraic relationship between those quantities.

II.7 The inflection point problem for plane curves of order three

Finally, we need to point out some applications and generalizations of invariant theory, for which there was no appropriate place so far. In particular, we have up to now said hardly anything about the geometric applications, which are especially far reaching. Here too we have to be satisfied with an example. From the extensive material we select the well-known inflection point problem for plane curves of order three.

Suppose given a ternary cubic form:

$$f(x_1, x_2, x_3) = a_{111} x_1^3 + \cdots + a_{333} x_3^3,$$

written symbolically as

$$\begin{aligned} f &= (a_1 x_1 + a_2 x_2 + a_3 x_3)^3 = (b_1 x_1 + b_2 x_2 + b_3 x_3)^3 \\ &= (c_1 x_1 + c_2 x_2 + c_3 x_3)^3 = (d_1 x_1 + d_2 x_2 + d_3 x_3)^3 . \end{aligned}$$

The latter expression is to be understood as follows. One only needs to calculate the exponents with the Binomial Theorem; this way one obtains exclusively terms of the form

$$c \cdot a_i a_k a_l x_i x_k x_l \quad (\text{respectively} \quad c \cdot b_i b_k b_l x_i x_k x_l, \ldots).$$

Into this one has to substitute everywhere a_{ikl} for $a_i a_k a_l$; likewise, one has to substitute a_{ikl} for $b_i b_k b_l$, etc. Incidentally, this is also the way in which one has to choose the multinomial coefficients in the explicit expression for f.

The equation $f = 0$ represents a plane curve of order three. An easy calculation shows that f has the following covariant:

$$h = \begin{vmatrix} \frac{\partial^2 f}{\partial x_1^2} & \frac{\partial^2 f}{\partial x_1 \partial x_2} & \frac{\partial^2 f}{\partial x_1 \partial x_3} \\ \frac{\partial^2 f}{\partial x_2 \partial x_1} & \frac{\partial^2 f}{\partial x_2^2} & \frac{\partial^2 f}{\partial x_2 \partial x_3} \\ \frac{\partial^2 f}{\partial x_3 \partial x_1} & \frac{\partial^2 f}{\partial x_3 \partial x_2} & \frac{\partial^2 f}{\partial x_3^2} \end{vmatrix}.$$

This is the so-called *Hessian covariant.* The equation $h = 0$ represents a new curve of order three, which, therefore, has an invariant relationship to the curve $f = 0$. The curve $h = 0$ is called the *Hessian curve* of the curve $f = 0$. If the coefficients of f are generic, then the curve $f = 0$ has nine inflection points, which are precisely the nine points of intersection of the two curves $f = 0$ and $h = 0$. If the coefficients of f are real, then exactly three of these inflection points are real. The nine inflection points can be found algebraically as follows.

For ternary forms, we have—in analogy to binary forms—the theorem that every homogeneous integral rational function of determinants of the type

$$\begin{vmatrix} a_1 & b_1 & c_1 \\ a_2 & b_2 & c_2 \\ a_3 & b_3 & c_3 \end{vmatrix},$$

viewed symbolically in a suitable way, is an invariant. Then the form f has the following invariant:

$$\mathcal{S} = \begin{vmatrix} a_1 & b_1 & c_1 \\ a_2 & b_2 & c_2 \\ a_3 & b_3 & c_3 \end{vmatrix} \begin{vmatrix} a_1 & b_1 & d_1 \\ a_2 & b_2 & d_2 \\ a_3 & b_3 & d_3 \end{vmatrix} \begin{vmatrix} a_1 & c_1 & d_1 \\ a_2 & c_2 & d_2 \\ a_3 & c_3 & d_3 \end{vmatrix} \begin{vmatrix} b_1 & c_1 & d_1 \\ b_2 & c_2 & d_2 \\ b_3 & c_3 & d_3 \end{vmatrix}.$$

To obtain its explicit expression one has to expand this expression and make the above substitutions. The invariant $\mathcal{S}$ has degree four. One can furthermore prove the theorem—which we will not do here—that except for f and h, f has essentially no further covariants of order three. This theorem is important for what follows.

We consider the family of curves

$$\alpha f + 6\beta h = 0.$$

All curves of this family pass through the nine inflection points of $f = 0$. The Hessian curve of a curve of the family, $\mathcal{H}(\alpha f + 6\beta h) = 0$, must again be a curve in the family, since $\mathcal{H}(\alpha f + 6\beta h)$ is a covariant of order three of f, hence must be of the form

$$\mathcal{H}(\alpha f + 6\beta h) = Af + Bh.$$

Thus it follows that all curves in the family $\alpha f + 6\beta h = 0$ also have the same inflection points as $f = 0$ (the inflectional tangents are, of course, different). If one calculates A and B, then one finds

$$A = 2\alpha^2\beta\mathcal{S} + 8\beta^3\mathcal{S}^2 + \alpha\beta^2\mathcal{J},$$
$$B = \alpha^3 - 2\beta^3\mathcal{J} - 12\alpha\beta^2\mathcal{S},$$

where the coefficients again have to be invariants. Accordingly, we know two invariants of f, namely, $\mathcal{S}$ and $\mathcal{J}$; these form, as one can prove, the full invariant system of the ternary cubic form. Incidentally, the form $\mathcal{J}$ is of degree six in the a_{ikl}.

We ask: Are there curves in the family that are their own Hessian? To answer this question, we need to determine α, β (that is, their ratio) such that

$$\mathcal{H}(\alpha f + 6\beta h) = C \cdot (\alpha f + 6\beta h);$$

so we must have

$$A = C\alpha = 2\alpha^2\beta\mathcal{S} + 8\beta^3\mathcal{S}^2 + \alpha\beta^2\mathcal{J},$$
$$B = 6C\beta = \alpha^3 - 2\beta^3\mathcal{J} - 12\alpha\beta^2\mathcal{S},$$

and after eliminating C,

$$\alpha^4 - 24\mathcal{S}\alpha^2\beta^2 - 8\mathcal{J}\alpha\beta^3 - 48\mathcal{S}^2\beta^4 = 0.$$

This equation has four distinct roots for $\frac{\alpha}{\beta}$. Thus, there are four such curves. Each of these curves must pass through the nine inflection points. Each of them must furthermore intersect itself in its inflection points; that is, the curves should have inflection points everywhere. But the only possible interpretation of this is that they are made up of straight lines; thus, each of them decomposes into three straight lines. Since a straight line can intersect a curve of order three only in three points, each of these twelve lines must contain three inflection points, and all nine inflection points lie on the circumference of every "syzygetic triangle." Thus, we have the following:

Theorem *The nine inflection points of a curve of order three (without double points and cusps) lie on twelve lines in groups of three, such that through each inflection point there pass four lines. These twelve lines can* *be ordered into four groups of three each so that each group contains all inflection points.*

The analytical determination of the inflection points poses no more difficulties from this point on. One determines the four values of $\alpha : \beta$; this requires the solution of an equation of degree four. Then, one decomposes each of the four forms $\alpha f + 6\beta h$ into linear factors, which must be possible. One sets these linear forms all equal to zero; then, one finds that each pair of these equations always has a common root such that one obtains a total of nine distinct such roots, which occur more than once (the others are to be omitted). These roots give the inflection points.

The equation for $\frac{\alpha}{\beta}$ has another special feature. For this biquadratic form in α, β, the invariant is

$$\begin{aligned} i &= a_0 a_4 - 4a_1 a_3 + 3a_2^2 \\ &= -48\mathcal{S}^2 + 3 \cdot 16\mathcal{S}^2 \\ &= 0. \end{aligned}$$

So it follows that: *The four syzygetic triangles have equiharmonic positions with respect to each other.*

Lecture L (August 5, 1897)

II.8 The generalization of invariant theory

The concept of an invariant is capable of generalization, as we have indicated earlier. The infinitely many linear transformations form a group, the most general group of linear transformations. We now select from this most general group, the so-called *projective group*—a certain system of linear transformations, which form a group. The group property, which was self-evident earlier, is now made an explicit hypothesis. The important problem of determining all relevant subgroups of the projective group has been completely solved by Sophus Lie through his theories, and this solution is one of the main applications of his theories. For such a group of linear transformations one can formulate the concept of an invariant just as in the general case: One is supposed to take a general form, transform it via an arbitrary linear transformation in the subgroup, and ask for those integral rational functions in the coefficients of the form that take on the same values, up to a multiplicative power of the substitution determinant, for the given and the transformed form. One can now pose the same questions for these invariants, which of course always contain the general invariants, that were considered for general transformations.

One of the simplest examples, and at the same time one of the most important, is that of the orthogonal substitution, for instance, for three variables. One calls *orthogonal* those substitutions whose application transforms the sum of the squares of the variables into itself, up to a constant factor. The form

$$f = x_1^2 + x_2^2 + x_3^2$$

will therefore be transformed into

$$f = f' = c\left(x_1'^2 + x_2'^2 + x_3'^2\right).$$

It is evident that the orthogonal substitutions form a group; since, if one transforms $\frac{1}{c}f'$ further into

$$\frac{1}{c}f = \frac{1}{c}f' = f'' = c'\left(x_1''^2 + x_2''^2 + x_3''^2\right),$$

then we also have

$$x_1^2 + x_2^2 + x_3^2 = c''\left(x_1''^2 + x_2''^2 + x_3''^2\right),$$

which precisely expresses the group property.

A general theorem now says that if one chooses an irreducible subgroup from a group, then this subgroup admits a rational representation through parameters. The orthogonal substitution—always for three variables—allows a rational representation through four parameters, since its general form, as one can show, is

$$\begin{aligned}
x_1 &= \left(a_1^2 + a_2^2 - a_3^2 - a_4^2\right) x_1' - 2\left(a_1a_3 + a_2a_4\right) x_2' - 2\left(a_1a_4 - a_2a_3\right) x_3',\\
x_2 &= 2\left(a_1a_3 - a_2a_4\right) x_1' + \left(a_1^2 - a_2^2 - a_3^2 + a_4^2\right) x_2' - 2\left(a_1a_2 + a_3a_4\right) x_3',\\
x_3 &= 2\left(a_1a_4 + a_2a_3\right) x_1' + 2\left(a_1a_2 - a_3a_4\right) x_2' + \left(a_1^2 - a_2^2 + a_3^2 - a_4^2\right) x_3'.
\end{aligned}$$

Here, a_1, a_2, a_3, a_4 are arbitrary parameters. Thus, this group is a four-parameter subgroup.

Now the main task is to determine all invariants of an arbitrary ternary form

$$f = A_1x_1^n + \cdots + A_Nx_3^n$$

with respect to an orthogonal substitution. Such invariants are, first of all, the general invariants of f. Secondly, however, we certainly also get all simultaneous invariants of f and the form

$$x_1^2 + x_2^2 + x_3^2$$

as invariants of f. That is, we need to determine all simultaneous invariants of f and the general ternary quadratic form $f_1 = \sum b_{ik} x^i x^k$. Let $\mathcal{S}(A, b)$ be such an invariant; then we certainly also get

$$\mathcal{S}(A', b') = \delta^p \mathcal{S}(A, b)$$

for the orthogonal substitution. If, in particular, one sets $b_{ii} = 1,\ b_{ik} = 0$ if $i \neq k$, then, according to our hypothesis, we get $b'_{11} = b'_{22} = b'_{33} = c,\ b'_{ik} = 0\ (i \neq k)$. But, since $\mathcal{S}$ is homogeneous in the b, it follows that

$$\mathcal{S}(A', b') = c^{p_1} \cdot \mathcal{S}(A', 1),$$

hence

$$c \cdot \mathcal{S}(A', 1) = \delta^{p_2} \cdot \mathcal{S}(A, 1),$$

so $\mathcal{S}(A, 1) = \mathcal{F}(A)$ is indeed an invariant. Incidentally, this argument is valid in general.

Theorem *Suppose there is a group of linear transformations such that each of them leaves a given form invariant; then the invariants with respect to this transformation group always include the invariants of the form f and the simultaneous invariants of this form and those given forms which are left invariant.*

This generalization, therefore, utilizes our previous theories. The question arises now whether, conversely, one obtains all invariants in this way. While it is probable, it has not been proven yet, even for orthogonal substitutions. This would completely answer the question about the finiteness of the full invariant system.

However, one can prove the finiteness of the full invariant system for the orthogonal substitutions of three variables completely analogously to the proof for the general invariants (cf. Hilbert 1890). To see this, one only needs to give a process Ω, which creates invariants. But here, such a process is

$$\Omega = \frac{\partial^2}{\partial a_1^2} + \frac{\partial^2}{\partial a_2^2} + \frac{\partial^2}{\partial a_3^2} + \frac{\partial^2}{\partial a_4^2}.$$

One has the following:

Theorem *If one transforms a system of base forms orthogonally and* *takes an arbitrary integral rational homogeneous function of the transformed coefficients, multiplies it with an arbitrary power, with a non-negative integer exponent, of* $(a_1^2 + a_2^2 + a_3^2 + a_4^2)$, *and applies to this function the process* Ω *enough times to result in an expression free of* a_1, a_2, a_3, a_4, *then the latter is always an orthogonal invariant.*

This theorem then implies easily the finiteness of the full invariant system; and because of this fact, one then proceeds immediately to all the observations we made about invariant fields, etc.

Also, Hurwitz has recently shown that all these theories of Hilbert can also be generalized to the orthogonal substitutions with arbitrarily many variables.

Let us mention another special invariant theory. Suppose we are given a space curve of order three. One can ask about all those transformations that leave this space curve unchanged. These, of course, form a group, namely—if we focus on a particular example (cf. Hilbert 1890, p. 533)—a four-parameter group of quaternary substitutions. In this case too, there is a process Ω, analogous to the previous one, so that everything procedes as before. At the same time, there exists a certain quaternary form $\mathcal{F}$ of order four that remains unchanged under these substitutions. This form is the discriminant of the cubic equation in t:

$$x_1 t^3 + x_2 t^2 + x_3 t + x_4 = 0.$$

Here too, the question arises whether the invariants consist of the general invariants of f and those of f and $\mathcal{F}$.

The two subgroups mentioned here play a special role in geometry. Even more important, however, are the translation and the rotation groups, and more important yet is the group of motions. The latter is given by a system of linear transformations, which form a group. It is characterized by the property that the transformations leave the two imaginary circle points and the line at infinity fixed. If this line is x_3, then the motion leaves

$$x_3 \qquad \text{and} \qquad x_1^2 + x_2^2 + x_3^2$$

fixed. But it suffices to consider only one form if one introduces line coordinates u_1, u_2, u_3 instead of point coordinates; in invariant theory, these are also called contragredient coordinates—we will return to these. The motion is then simply characterized by the property that the form

$$u_1^2 + u_2^2$$

remains unchanged. This leads us to consider other coordinates besides the ordinary point coordinates, which we want to do now, at the end.

In order to return once more to the group of the motions, one first needs to transform a form $f(x_1, x_2, x_3)$ into contragredient coordinates $\mathcal{F}(u_1, u_2, u_3)$, and then ask about the invariants which arise under the group of substitutions that leave the form $u_1^2 + u_2^2$ unchanged. For two straight lines, we obtain in this way only the cosine and the sine of the angle they form; for a triangle, certainly the sine and cosine of the angles, the length of the sides, the area, etc., are invariants. Presumably, here too, the theorem that all invariants of the form $\mathcal{F}$ with respect to the motion of the plane are simultaneous invariants of $\mathcal{F}$ and $\Phi = u_1^2 + u_2^2$ remains valid; the converse is self-evident.

Analogous observations hold for three-dimensional space. Here, the group of transformations is characterized by the property that the form

$$u_1^2 + u_2^2 + u_3^2$$

remains unchanged.

Lecture LI (August 6, 1897)

II.9 Observations about new types of coordinates

In this final section we want to point out briefly some further generalizations of invariant theory, namely, some new types of coordinates.

Let f_1, f_2 be two ternary forms of the same order. The forms

$$\kappa_1 f_1 + \kappa_2 f_2$$

form a so-called sheaf of forms. Here one can view κ_1, κ_2 as binary variables and subject them to a linear transformation that is completely independent of x_1, x_2, x_3, the arguments of f_1, f_2. In this way, one is led to forms with several sequences of variables, and can now, just as before, ask about the invariants of these forms. There already have been extensive investigations in this respect, but the subject has not yet been completely covered. Here too, one obtains differential equations that are analogous to the earlier ones, and in many respects one returns to the old invariant theory. In general, one of course has to be concerned about using the earlier theorems as much as possible to obtain the generalizations. One obtains the invariants by successively first considering one, then two, ... sequences of variables as transformation variables.

For applications to geometry, however, it is not sufficient to consider only the case where the individual sequences of variables are subjected to distinct transformations. While one often has to deal with several sequences of variables, these will be subjected to transformations which are dependent on each other. Here too, the invariants are defined as before. But their study has not yet been carried out in general. We limit ourselves to point out the specific problems so far as they have geometric significance. We distinguish two cases.

First, assume that the number of variables in the variable sequences is always the same, and they are all subjected to the same linear transformation, for instance, in the form $f(x_1, x_2, x_3;\ y_1, y_2, y_3)$, where, of course, the x and y are independent of each other. After Sylvester and Cayley, such variable sequences are called *cogredient variable sequences.* These occur in geometry, namely, during the formation of polars.

Second, the individual transformations may depend on each other in a general way. In the plane one has to consider two sequences of variables: the point coordinates x_1, x_2, x_3, and the line coordinates u_1, u_2, u_3. The transformations of the latter are clearly determined once those of the former ones are determined. One obtains the relationship by observing that the condition of the relative position of points and lines

$$u_1x_1 + u_2x_2 + u_3x_3 = 0$$

has to be preserved by the transformation. If

$$\begin{aligned} x_1 &= \alpha_{11}x_1' + \alpha_{12}x_2' + \alpha_{13}x_3', \\ x_2 &= \alpha_{21}x_1' + \alpha_{22}x_2' + \alpha_{23}x_3', \\ x_3 &= \alpha_{31}x_1' + \alpha_{32}x_2' + \alpha_{33}x_3', \end{aligned}$$

then we must have

$$\begin{aligned} &(\alpha_{11}x_1' + \alpha_{12}x_2' + \alpha_{13}x_3')\,u_1 + (\alpha_{21}x_1' + \cdots)\,u_2 + (\alpha_{31}x_1' + \cdots)\,u_3 \\ &\qquad = u_1'x_1' + u_2'x_2' + u_3'x_3'. \end{aligned}$$

From this it follows that

$$\begin{aligned} u_1' &= \alpha_{11}u_1 + \alpha_{21}u_2 + \alpha_{31}u_3, \\ u_2' &= \alpha_{12}u_1 + \alpha_{22}u_2 + \alpha_{32}u_3, \\ u_3' &= \alpha_{13}u_1 + \alpha_{23}u_2 + \alpha_{33}u_3, \end{aligned}$$

which makes it easy to see how the u can be expressed in terms of the u'. Suppose now that we are given a form $f(x_1, x_2, x_3) = a_{111}x_1^3 + \cdots$.

Aside from the in- and covariants, one can inquire about those forms $\mathcal{F}(a, u_1, u_2, u_3)$ for which

$$\mathcal{F}(a', u') = \delta^p \mathcal{F}(a, u).$$

These forms are called *contravariants.* Such a contravariant is clearly the left side of the equation of the curve $f = 0$ in line coordinates. If a form $\mathcal{F}(a, x, u)$, which contains all three groups of variables a, x, u, has that property, then it is called an *intermediate form.* The variables u are called *contragredient variables.* One can, of course, also consider simultaneous form systems. For instance, to determine the invariants of a form f with respect to a motion, one has to determine the simultaneous invariants of $f(x_1, x_2, x_3)$ and $\phi - u_1^2 + u_2^2$.

Analogous observations hold in space, that is, for the quaternary case. Aside from the point coordinates, one also has to consider the plane coordinates u_1, u_2, u_3, u_4 in this case. If one forms the determinant

$$\begin{vmatrix} x_1 & x_2 & x_3 & x_4 \\ y_1 & y_2 & y_3 & y_4 \\ z_1 & z_2 & z_3 & z_4 \\ t_1 & t_2 & t_3 & t_4 \end{vmatrix},$$

where the x, y, z, t are all subjected to the same linear transformation, namely, that of the point coordinates, then the plane coordinates u_1, u_2, u_3, u_4 are transformed exactly as the minors complementary to x_1, x_2, x_3, x_4, respectively. In space this fact gives rise to another observation. Namely, we now determine the minors of order two, that is, the quantities

$$p_{ik} = x_i y_k - x_k y_i,$$

and observe how these are transformed. These p_{ik}, of which there are six, are very important; they characterize a line in space, because of which they are called line coordinates. Between them exists the following relation:

$$p_{12}p_{34} + p_{13}p_{42} + p_{14}p_{23} = 0.$$

One arrives in this way at a new type of invariant object, namely, that which contains the p and may or may not contain the x and u. This is the basis for Plücker's *line and axis geometry*; he was the first to investigate the p more closely. The generalization to even more variables is evident.

Modern geometry has recognized that it is a one-sided and unjustifiable specialization to view the point as the element on which everything is based, that is, to interpret only forms of three or four variables geometrically. One is then led geometrically and analytically to great difficulties, which are not inherent in the situation. One thus must introduce further coordinates. The oldest idea is to use the line as the new element in the plane, and the plane as the new element in space. One can then build a geometry just as before, and now the point becomes what in point geometry was the line (respectively the plane). This constitutes the principle of duality. But an essential extension occurs if one uses the straight line as the basis in space, as Plücker did. The straight line in space is given through six homogeneous variables p_{ik}, between which there exists a quadratic relation; thus, there is a four-fold infinity of lines in space. A linear equation between the p_{ik}

$$a_{12}p_{12} + \cdots + a_{34}p_{34} = 0$$

constitutes a three-fold infinite family of lines; this family is called a *linear complex*. Similarly, one can establish arbitrary equations between the p and interpret them geometrically, or one can consider several equations simultaneously. In this way, one arrives at the ruled surfaces, etc. The situation is analogous in spaces of higher dimensions.

An equation $f(p_{ik}) = 0$ always represents a three-fold infinite family of lines, a "complex." The in-, co-, contravariants, etc., can still be defined in the same fashion. An essential property of this line geometry is that in it points and lines play the same role. Therefore, everything becomes very general and elegant.

One can carry the generalization even further. One can take any geometric object or system of algebraic forms as the basic object, and can then raise all the questions of ordinary geometry. Certain systems of equations represent geometric objects, such as point, plane, etc.

The equation

$$axy + bx + cy + d = 0,$$

for instance, represents a three-fold infinite system of hyperbolas in ordinary Cartesian geometry. A homogeneous relation between the coefficients $\mathcal{F}(a, b, c, d) = 0$ cuts out a two-fold infinite family of hyperbolas. A point is given by a relation that says the hyperbolas with coefficients satisfying this relation pass through the point.

Another example is that of a pair of points in the plane. There are four-fold infinitely many such pairs of points. A point is given through the relation between the coefficients of the pair of points that says that the two points coincide.

We get a new geometry if we use a line and a point on it as the basic element. This object is three dimensional. One calls such a system a *connex*. The connexes are very important.

Finally, a very well-known and much studied geometry is that which is based on the circle in the plane, and the sphere in space. The latter is called *spherical geometry*. The sphere is given by the equation

$$k_1\left(x^2+y^2+z^2\right)+k_2x+k_3y+k_4z+k_5=0.$$

Thus, spherical geometry is four-dimensional. One can ask about the significance of the linear complexes. A point is determined through that quadratic relation between the k that says that the radius of the sphere is zero. The plane simply has the equation

$$k_1=0.$$

It is clear how one can raise further questions and treat them geometrically or invariant theoretically.